Boeing 727 Study Guide

By Rick Townsend

This book was originally developed for pilots at American Airlines. Neither American Airlines nor Boeing Company have endorsed or in any way been responsible for the content of this book.

As a courtesy to the pilots of American Airlines who have supported the development of these books, it is noted that this edition is current to the best ability of the author to make it so.

Published by Pilot Study Guides, LLC, Grapevine, Texas.
First Edition and Copyright 1992
Updated continuously (including copyright protection) through the current 2019 edition.

Graphics: With one exception, all graphics were produced by and are the copyright-protected property of the author, except where noted within this edition.

Acknowledgement: Thanks to Norebbo Design Studios, _www.norebbo.com_ for allowing use of the profile drawing in the Airplane General section (planned for a later edition).

Corporation References: American Airlines™ and the Boeing Corporation™, referenced herein, have neither endorsed nor provided any proprietary information for use in these books. Their referencing is for informational awareness only.

ISBN: **978-1-946544-30-8** (Print)
ISBN: **978-1-946544-24-7** (Amazon Kindle)

Table of Contents

Introduction

What This Book IS

This Study Guide is a compilation of notes taken from the flight manual, class notes, PLATO, and operational experience. It is intended for use by new hires and initial qualification crewmembers preparing for orals, and also for systems reviews prior to recurrent training or check rides. It is assembled in an attempt to organize in one location all the buzz words, acronyms, and numbers the average guy needs to know in order to get through the events above from an aircraft systems standpoint.

What this Book IS NOT

It is not officially sanctioned by American Airlines, and the author assumes all responsibility for accuracy. (Forward corrections to the address below, please!) It does not replace study of the operations manual, but instead provides a supplementary source of review material to complement study of official publications. It does not include Flight Manual Part 1, or other materials--just airplane systems and some ideas for in flight data-tracking. The Study Guide is not printed on fancy paper or expensively bound. That makes it easy to take apart and put in your flight manual with the appropriate sections, or keep together as you see fit. Therefore, for these same reasons, it is also not expensive. It is intended to be affordable and usable, instead.

Suggested Uses

A good time to use this book is on layovers. A reasonable plan would be to attempt to review limitations, Red Box and Black Border items once per sequence, and the rest of the sections once per month. Those pages (1-10) can be kept on a clipboard for ready reference in the cockpit, as well. Reviewing sectional study outlines usually leads to trying to remember one of those great plumbing diagrams from the flight manual, and so that also gets reviewed in the process.

Limitations, Red Box and Black Border items are laid out so that you may cover the answer with a 3x5 card and quiz yourself. Other sections are in outline form.

A word about acronyms--You will probably see most of the acronyms you have heard of before in this booklet, as well as a few new ones. Most of us avoid them when possible, using clues on aircraft instrument and systems management panels to jog our memories. But for those you have trouble with, I've included all the ones I could think of. It is not necessary to learn all of them, but only those you need to help remember tough sets of items.

The Flight Data Forms provided have their own explanatory notes in the last chapter. They provide an excellent way to record data for logging later. I do this monthly on my computer, and have the paper copy for validation purposes. They are intended for duplication for the personal use of folks who buy this study guide. This is an inexpensive way to keep a permanent record of OUT, OFF, ON and IN times, delay codes, and so on. So when you get that call later in the month requesting information, you'll be prepared!

This material is copyrighted. Please don't reprod**2019**ook, as you'll soon be able to see when you study the material. Flight Data Sheets, anti-icing procedures and related instructions (pages 45-46) may be copied by customers without restriction. Thanks.

Corrections and Inputs

This book was compiled when I was flying the Boeing 727, and the fleet at my airline left the scene many years ago. I can't guarantee the accuracy of the contents with any current guidelines from the manufacturer. Use at your own risk! Let me know of any suggested changes, please.

Page References

All references to operating manual pages were from the manuals as they existed at the time of compilation. These manuals no longer exist (to the author's knowledge) and thus page references should be taken as representative regarding where information may be found in legacy manuals the user may possess.

Rick Townsend............................rick@pilotstudyguide.com
www.pilotstudyguide.com

© 1992 by Rick Townsend   Updated 1993-2019

2 -- Limitations

General & Operational Limitations

Temperature Limits	-54°C to +49°C
Max Altitude	8300 Feet **Note:** *When pressure altitude is below -1000', use data for -1000'*
Runway Slope Limit	+2%

T/O & Landing--Wind Limits (Including gusts)	
Runway Dry (Max Demonstrated):	29 Knots
Tailwind (AFM):	10 Knots (May be further limited by performance requirements)

Visibility Less than ¾ Mile (RVR 4000' or 1200 Meters), But Not Less Than ½ Mile (RVR 1800' or 550 Meters):	15 Knots (NORMALS, p. App-Ldg-G/A, Page 15)
Less Than ½ Mile (RVR 1800 or 550 Meters):	N/A (Cat II approaches not authorized per PB 273)

Note: *The remainder of these crosswind limits are found in areas other than the Ops Manual Limitations section and are consolidated here.*

Braking Action (AA Policy) Fair:	20 Knots	*Braking action must be determined by ATIS, PIREPs, tower*
Poor:	10 Knots	*reports or environmental conditions including wet runway, standing water, slush, snow or ice (Flt. Man. Pt. I, page 10-20)*
Maximum Wind Gust:	50 Knots, except in emergency (FM Part 1, Section 10, page 20)*	
Runway Width Less than Standard:	20 Knots (FM Part 1, Section 10, page 20)*	
More than 1 LED Not Extended:	19 Knots (Section 4, page 31)	
Anti-Skid INOP:	15 Knots (Section 16, page 11, and MEL System 32)	
Rudder Low Pressure:	10 Knots (Section 12, page 16)	

Maximum Operating Altitude	42,000 Feet

Airspeed Limits

Mode A, and All A/C not Dual-V_{MO}		V_{MO}	M_{MO}	
Equipped:	Sea Level:	380 Kts	----	**Note:** *For limitations involving other altitudes, see LIM p. 2.*
	21,500' MSL:	411 Kts	.90	
	Over 21,500' MSL:	----	.90	

Mode B Definition:	♦ In Flight Gross Weight over 172,000 pounds			
	♦ Zero Fuel Weight over 136,000 pounds			
Limitations:		V_{MO}	M_{MO}	
	Sea Level:	350 Kts	----	**Note:** *For limitations involving other altitudes, see LIM p. 2*
	26,500' MSL:	372 Kts	.90	
	Over 26,500' MSL:	----	.90	

Landing Gear: Extending:	270 Kts or .83 Mach
Extended (V_{LE}/M_{LE}):	320 Kts or .83 Mach
Retracting:	200 Kts (**Note:** *Monitor flap speed and gross weight schedule for heavy-weight retraction after takeoff if gear are to be recycled.*)

Leading Edge Devices Extended (One or more)	240 Kts

Wing Flaps, Operating and Extended 2:	230 Knots
5:	215 Knots
15:	205 Knots
25:	185 Knots
30:	180 Knots
40:	N/A

One or both Yaw Dampers Inop: >FL 300:	Normal Operations Not Permitted
If failure occurs, >FL 300':	Descend at 0.8 Mach to 30,000' or below

FL 300 & Below:	270 Kts
FL 250 & Below:	350 Kts
Maximum Tire Ground Speed	182 Knots
Fuel Dumping	Same as V_{MO}/M_{MO}

Weights

		-200A	-200B	
Maximum Taxi Weight		178,500	191,000	
Maximum Takeoff Weight		177,900	190,500	(See notes below.)
Max in flight Weight, (AFM)	Flaps 25°:		185,800	
	Flaps 30°:	155,500	155,500	
Max Landing Weight, (AFM)	Flaps 30°:	154,500	154,500	

Note: *Mechanical stop prevents 40 ° flap selection.*

Maximum Zero Fuel Weight (AFM)		138,000	140,000

Note 1:: *See Section 2, Page 3 for miscellaneous limits involving other flap settings and elevations.*

Note 2: *The maximum Zero Fuel Weight is a wing structural limit. All weight in excess of the Maximum Zero Fuel Weight must be fuel.*

Landing Over the Maximum Weight	For all B-727 models, landing at weights greater than the Maximum Landing Weight (AFM) is authorized. Refer to Part 1 for overweight landing policy, and to the Overweight Landing procedure in Section 4 and/or the Performance Manual to determine the maximum allowable overweight landing weight and for overweight landing procedure guidelines.

Note 3: *Maximum landing weight exceedance requires OF-25 and E-6 entires.*

Air Conditioning & Pressurization

Ground-Supplied Conditioned Air		
	Max Temperature:	60 C/140°F
	Max Air Flow:	250 Lbs/Minute
Max Differential Pressure (AFM)		8.6 ± 0.15 psi
System Safety Differential Pressure (AFM):		9.6 psi Maximum
Maximum Differential Pressure for T/O & Landing (AFM):		0.125 psi
Unpressurized Flight--Number of Packs Required:		Only one pack may be on for unpressurized flight.
Max altitude for unpressurized flight following an in-flight depressurization:		14,000'--But may be exceeded where terrain clearance requirements dictate.
Maximum altitude when the aircraft is dispatched for unpressurized flight:		10,000'
Engine #2 Bleed Restriction (AFM)		**Caution:** *When operating at Maximum Takeoff Thrust or Maximum Continuous Thrust, engine 2 shall not be used as a bleed air source for air conditioning until flap retraction is complete and all obstacles are cleared.*

Anti-Icing & Rain Protection

Engine Anti-Ice is Required (AFM):	-For all ground or flight operations where icing is encountered or anticipated. *Exception*--During **climb** and **cruise** with temperature below -40° SAT; **Descent** through these conditions still requires anti ice. **Note**: *See Section 6 of this Study Guide for Icing Condition definition.*
In Flight Minimum N_1 **In icing conditions:** **(AFM)** **In moderate to severe conditions:**	55% 70% N_1 or higher, except as required for landing. **Note**: *Moderate to severe is defined as with TAT less than -6.5°C or 20°F*
Max Temperature for Wing Anti-Ice	+10°C TAT (Operation above this temperature can reduce hail-resistance characteristics of the leading edge.)
Window Heat **When required on:**	10 minutes prior to takeoff--Windows 1 & 2
Limitations if INOP to any window:	A/S limited to 250 Kts below 10,000', or higher if birds possible
Standing Water/Slush Requirements	♦ Chine Tires Installed (*Always*) ♦ Deflectors on Main Gear (*Always*) ♦ Static Heaters operative with temperature below 2°C

Autopilot

Maximum A/S $\quad\quad\quad\quad\quad\quad\quad\quad\quad\quad\quad$ V_{MO}/M_{MO}

Minimum Altitude Enroute, Including Climb & Descent (Excluding Approaches):	1100 Ft (AFL)

Approaches	
ILS Coupled Approach, IFR:	80 Feet AFL
ILS Coupled Approach, VFR:	50 Feet AFL

Engine-Out Autopilot-Engaged Approaches: $\quad$ Autopilot use not authorized for either precision or non-precision approaches; must be disengaged prior to commencing final approach segment.

Manual G/S Mode: $\quad$ Use prohibited by American Airlines policy.

APU

In Flight Use (AFM) $\quad\quad\quad\quad\quad\quad\quad\quad$ Not allowed

Refueling $\quad\quad\quad\quad\quad$ **Over Wing:** $\quad$ No APU use if over wing refueling filler caps not installed

$\quad\quad\quad\quad\quad\quad\quad\quad$ **Running APU:** $\quad$ Do not start or stop APU during refueling operation. An explosion can occur.

EGT $\quad\quad\quad\quad\quad\quad\quad\quad\quad\quad\quad$ 710°C (Shut down APU if this value is exceeded.)

Starter Motor Duty Cycle $\quad\quad\quad\quad$ 1 minute on, 4 minutes off.

APU Generator or External Power	
Frequency	400 ± 10 Hz
Voltage	115 ± 5 Volts
Max Continuous Load	165 Amps (*APU Only, not an External Power Limit*)

Electrical Power

A/C Generator	
Frequency	400 ± 8 Hz
Max Continuous Load	36 KW
Max Load for 5 Minutes Max	54 KW
Voltage	115 ± 5 Volts

Maximum Total Load for 2-Generator Operations	54 KW (*Section 10, p. 28*)

AC Load Division--Difference with Generators Operating in Parallel	
Real Load	7 KW
Reactive Load	3 KVAR

Transformer-Rectifier	
Min Voltage Under Load	24 Volts
Max Load per TR	50 Amps

Fire Protection

Engine Fire Bottles	570-690 psi at 70°F*
APU Fire Bottles	350-430 psi at 70°F *Note: *See Chart, p. 2-8 for other temperatures.*

Flight Controls

Stall Warning System	Must be operative for all normal flight operations.
Rudder and Aileron Trim	Maintenance action required prior to passenger flights if more than 2 units of aileron or rudder trim are required in normal cruise.
Handwheel--Maximum Allowable Coast	2 Revolutions
Speed Brake Operations with Wing Flaps (AFM):	Simultaneous deployment in flight is prohibited.

Flap Operations (AFM):

Max Landing Flaps, -200A:	25º (30º Reserved for emergency situations.)
2º Gate:	For Extension or Retraction In Flight, Handle must remain at 2º Gate until LED operation verified.
Maximum Altitude:	20,000' MSL

Alternate Flap Operations

Flight:	One full cycle (up and down), then 25 minutes off
Ground:	10 Minutes of operation, then 25 minutes off

Fuel

Minimum Tank Temperature	-37ºC
Maximum Tank Fuel Temperature	+49ºC
Fuel Freezing Temperature	About -40ºC
Outboard Tank Maximum Allowed Weight	12,638 Lbs.
Fuel Balance (Ground and Flight) *#2 vs. #s 1 & 3 Tank Limits:*	#1 or #3 quantity shall not exceed #2 quantity by more than 2000 lbs.
#1 vs. #3 Tank (AFM):	#1 & #3 Tank quantities must not differ by more than 1000 pounds.
Ballast Fuel	When loaded, cannot be burned

Fuel Usage, B-727 200B (AFM)

Aux Tank Maximum:	6106 Lbs.
T/O, Approach, Landing:	Aux tank use prohibited
T/O Gross Weights Above 178,000 Lbs., & When #2 Tank has More Fuel than #1 or #3:	Use tank #2 for all operations (including takeoff and landing)
T/O Gross Weights Below 178,000 Lbs., with NO Fuel in Aux Tank:	Use standard B-727 fuel burn procedures
Sequence for Using Aux Tank Fuel:	With fuel in aux tank, burn at least 8000 lbs. from tank #2 or until tanks #1 and 3 are equal before switching to aux tank. Fuel boost pumps in tanks #1, 2 and 3 must be on and all fuel manifold valves open before switching to aux tank fuel.
Aux Fuel Tank Use Monitoring:	Fuel feed from aux tank must be verified (engine indications normal and aux tank fuel quantity decreasing.)

See additional notes on –200B fuel usage, OM page LIM 2-7 & 2-8. See additional fuel system notes, next page of this Study Guide.

Additional Fuel System Notes:
1. 1000 Lbs limit does not apply at lower gross weights typical of maintenance operations
2. No structural inspection requirements pertain just to a lateral imbalance condition.
3. Any imbalance over the 1000 Lbs limit must be corrected prior to flight dispatch from the gate.

Hydraulics

Hydraulic Fluid Quantity

Minimum at Gate, System A	3.5 Gal.

Minimums for Takeoff	
System A	3.0 Gal
System B	Full
Standby	0.28 Gal.

Hydraulic Pressure

Minimum (No system actuated and interconnect closed):	2800 psi

Maximum Pressures

☐ Normal Operating Maximum:	3175 psi
☐ Maximum (System Relief Pressure):	3500 psi
☐ Airplane may not be dispatched with pump output over:	3175 psi
☐ In flight, with pump output between 3175 and 3500 psi:	Turn off affected pump (except for landing)
☐ Over Maximum:	Turn off any Pump with output over 3500 psi

B System Minimum Pressures	
Interconnect Closed, 1 B Pump ON:	2800 psi
Interconnect Open, Both B Pumps ON:	2800 psi
One B Pump ON:	2300 psi

Deferrable but Maintenance Required:	Below 2300 but not below 2100 psi, but only deferrable if at least 2800 psi with both pumps on
Maintenance Action Required for Flight:	One Pump ON and System Pressure below 2100 psi

Accumulator Pre-Charge Pressure	
Brake:	1000 ± 50 psi at 70ºF See Chart, Page LIM 2-9 for other temperatures..

Landing Gear

Brake Wear Limits	Brake assembly must be replaced when break wear indicator pins (with parking brake parked) indicate flush or below. Brake wear pins are not required to be installed. If missing, periodic check of brake wear by maintenance is required.

Pneumatic Brake Pressure	1200 ± 50 psi at 70ºF

Instruments

RMDI/CDI Indications	
CDI vs. Slaving RMDI:	±1º
Capt vs. F/O RMDI Card:	±5º
CA CDI course cursor vs. FO MDI #1 VOR Pointer:	±5º
FO CDI course cursor vs. CA RMDI #2 VOR Pointer:	±5º
#1 vs. #2 ADF/VOR Pointers tuned to same station:	±8º
Course Deviation Bar	The Navigation Section contains additional tolerances for CDI & HDI ground checks--See "VOR Ground Check," page 18-Nav-7.

TCAS (AFM) Compliance Requirements:	Initiating evasive maneuvers for TAs based only on information shown on the TCAS traffic display is prohibited. Compliance with an RA is required unless, in the opinion of the pilot flying, doing so would compromise the safe operation of the flight. See Limitation section, page 9 for other TCAS-related limits. Maneuvers in response to an RA which are in the opposite direction of that advisory are prohibited unless they are the only means to assure safe separation.

See additional TCAS notes, OM page LIM 2-9

Oxygen System

Oxygen System (70°F), Domestic & Extended Over-Water		
	Cockpit:	1200 psi @70°F
	Cabin:	1500 psi @70°F

Power Plant

Instrument Markings		
	Red Arc:	Prohibited Operating Range
	Red Radial:	Maximum and Minimum
	Yellow Arc:	Precautionary Range
	Green Arc:	Normal Operating Range

RPM

		-9A	-15A	Time
	N_1:	100.1%	102.4%	
	N_2:	100.0%	100.0%	

EGT

		-9A	-15A	Time
Starting;	≤59°F OAT:	350	575	Momentary
	>59°F OAT:	420	575	Momentary
Max Cruise Thrust:		525	550	Continuous
Max Continuous. & Norm Climb:		545	580	Continuous
Takeoff:		590	620	5 Minutes
Accel/Rapid Power App:		590	630	2 Minutes

Oil

Minimum Pressure:	40-55 psi;	(Refer to Power Plant abnormals if low)
Maximum Pressure:	55 psi;	(E-6 Entry if pressure above)
Temperature, (-9A):	120°C, 121-157°C Max for 15 Minutes	
(-15A):	130°C, 131-165°C Max for 15 Minutes	
Minimum Quantity Required for Dispatch:	4 Quarts+ 2 Quarts/hour for Planned Flight Time	
Maximum Consumption Rate Allowed:	1 Qt/hour Maximum	
Parking A/C, Duration of Next Flight Unknown:	2.5 Gallons (Sec 3, p. 46B)	

Reverse Thrust	In Flight:	Use Prohibited
	Static:	80%N_1, Max 10 Seconds, No Repeat before 3 Minutes
	Power Back:	80%N_2, Same as Static, Except 10 Second Limit N/A
	On Landing:	84% N_1 unless emergency dictates more power; F/E Calls 80% N_1 or "No reverser light No.___ engine." (p. 3-42)

Engine Ignition T/O & Landing (AFM):	Required to be ON

Engine Starter Duty Cycle	
First Attempted Start:	1 Minute ON, 1 Minute Off
Subsequent Starts:	1 Minute ON, 5 Minute Off
Motoring Engine with Fuel & Ignition OFF:	2 Minutes ON, 5 Minute Off

PDCS	Configuration Requirement:	Cannot use unless engine configuration displayed is the same as the engine configuration installed on the aircraft.
	One-Engine INOP Ferry Flights:	PDCS must be placarded inoperative.
	-200A, PDCS Use on Takeoff:	Not authorized

Miscellaneous Limits

Aft Stairs Considered an emergency exit; Will not normally be used for passenger emplaning/deplaning. ... *(See Limitations section, page 12 for additional comments.)*
Notes: *1. Use of aft stairs to deplane improperly boarded passengers is not authorized.*
2. The restriction on the use of the aft stairs applies to passengers only, and does not prohibit routine use by airline and other authorized personnel.
3. If the aft stairs are used for passenger emplaning or deplaning, refer to Miscellaneous Section—Abnormals.

Evacuation Slides
Arming Requirements—Airplane in Motion: Installed when passengers aboard for all Taxi, Takeoff, and Landing Operations
Prior to A/C movement; Passengers Aboard: At least one floor-level exit must be open or armed.
Pressure: 2700 psi @ 70°F

Dry Ice Limits **_Forward Cargo Compartment:_** 440 Lbs.
Aft Cargo Compartment: 1500 Lbs.
Passenger Cabin, Packages or Carry-on Baggage: 2 Kg (4.4 pounds) maximum, Twenty (20) pounds total, provided package permits release of carbon dioxide gas; OK-333 required.
Galleys: No restriction when carried in galleys.
Over 100 Lbs, A/C Parked: Respective Belly Compartment Door Open for ventilation
Over 300 Lbs on Board, Landing: With only 1 A/C pack operating, crew should use O_2 masks with regulators 100% for "3 minutes, about 20 minutes prior to landing."
Over 500 Lbs on Board, Flight: If A/C becomes unpressurized, one pilot must wear O_2 Mask until landing or A/C pressure is restored
Over 1000 Lbs on Board, Ground: ☐One A/C Pack must be kept running;
☐Ground Venturi Fan in GROUND VENTURI
☐Weather Permitting, keep one cockpit window open

Navigation Lights **_In Flight Use of ON BAT:_** Do not use.

Navigation

Global Nav./Flight Mgt. System (GFMS) **_Conditions of Approval:_** Approved with software version number HT9100-002A, (or later FAA approved version.)
Primary means of Navigation for Routes (Single System Installed): ♦Caribbean and Bermuda Routes
♦Within the Coterminous US & Alaska (below 70°N latitude)
Primary means of Navigation for Routes (Dual Systems Installed): Between 70°N and 58°S latitude for operations in oceanic and/or remote land areas
Continued Use, UNABLE RNP Displayed: ♦*En route and terminal IFR Operation can be continued providing the system position is verified every 15 minutes, using other approved navigation equipment.*
♦Crew must report degraded navigation performance to ATC.
Navigation in Dead Reckoning Mode: *Navigation* **cannot be predicated** *on this mode, but it may be used in the absence of any other means of navigation.*
Requirements Prior to Use: ♦*Navigation data base currency must be checked.*
♦*Use of out-of-date data base is prohibited unless each selected navaid and waypoint is checked for accuracy against current chart data*

Enhanced GPWS

Deviation from ATC Instructions: Authorized to extent needed to comply with Enhanced GPWS alert
EGPWS Must be Inhibited (OVRD) **_in Following Conditions:_** Before takeoff, or within 15 NM of approach or landing at an airport with
♦Longest runway less than 3500 feet in length *or*
♦No published approach procedure
Standby Altimeter Must be set to QNH (MSL) for system to operate
Instrument Panel Flood Lights: Must be operative if system is used at night
AA Policy on GFMS **_Holding Patterns:_** Use not authorized
Terminal area up to Final Approach: Use authorized up to but ***not including*** final approach segment.

4 -- Emergencies

Red Box Initial Action Items:

Explosive Depressurization *(4-9):*
O₂ MASKS .... ON
COMMUNICATIONS .... ESTABLISH
Initiate Emergency Descent.
PASSENGER OXYGEN SWITCH .... ON / CK LT ILLUMINATED

WARNING
The time of useful consciousness following an explosive decompression will vary from about 1-2 minutes at 25,000 feet to 15-23 seconds at 40,000 feet.

(Warning is included for study, not part of the red box procedure.)

Rapid Depressurization *(4-9):*
CARGO HEAT OUTFLOW SWITCH .... CLOSE

Loss of All Generators *(4-10):*
ESSENTIAL POWER SELECTOR .... STANDBY
BATTERY SWITCH .... CHECK ON
ALL FIELD RELAYS .... ATTEMPT to CLOSE /
CHECK VOLTS and FREQ

■ **If none of the field relays will close:**
BATTERY SWITCH .... CYCLE OFF / ON
ALL FIELD RELAYS .... ATTEMPT to CLOSE /
CHECK VOLTS and FREQ
ESSENTIAL POWER SELECTOR .... .TO AN OPERATING GEN

Engine Failure/Shutdown *(4-12):*
Engine Fire/Severe Damage/Separation *(4-11):*
Engine Fire on Ground *(4-12):*
ESSENTIAL POWER / KWS .... CHECK
Download as required.

Use of Cockpit Oxygen Masks and Smoke Goggles *(4-17):*
O₂ MASK and GOGGLES .... ON
COMMUNICATIONS .... ESTABLISH

Electrical Fire or Smoke - Unknown Source *(4-18):*
Consider landing at nearest suitable airport.
ALL FUEL BOOST PUMPS/CROSSFEED VALVES .... ON/OPEN
ALL BUS TIE BREAKERS .... TRIP
ALL GENERATOR BREAKERS .... TRIP

Runaway Stabilizer *(4-27):*
TRIM WHEEL .... STOP
STAB TRIM CUTOUT SWITCHES .... BOTH CUTOUT

GPWS Warnings *(MANEUVERS-6):*
Immediately and simultaneously:
AUTOPILOT .... DISCONNECT
THROTTLES .... FULL FORWARD
PITCH .... ROTATE TO 20° or GREATER
SPEED BRAKES .... RETRACTED

Windshear/Microburst Escape Procedure *(MANEUVERS-5):*
Announce *"Escape"* and –
Immediately and simultaneously:
AUTOPILOT .... DISCONNECT
THROTTLES .... FULL FORWARD
PITCH .... ROTATE TO 15°
SPEED BRAKES .... RETRACTED

Flight Engineer Initial Action Items:

Loss of APU Power *While Taxiing* *(10-26)*	*Brakes:*	Brake Interconnect	CHECK OPEN
	Packs:	Both Pack Switches	OFF
	Power:	No. 1 Gen Breaker	CLOSE
		KW's, Volts, and Frequency	Check

Fuel Dumping (14-Fuel-27) *8 Pumps ON:* Boost Pumps:
- No. 1 and No. 3 Tanks ....ALL ON
- No. 2 Tank (without aux fuel) ....ALL ON
- No. 2 Tank (with aux fuel) .... 2 ON
- Aux Tank (if required)..................ALL ON

Crossfeed Valves ....ALL OPEN
Dump and Nozzle Valves ....ALL OPEN

AC Power--General (10-25):

Essential: **Essential Power**..................AS REQUIRED
Check that the power supply is taken from a properly operating generator.

Reduce: ■ **If all generators are malfunctioning:**
Essential Power Selector ....STANDBY
Reduce heavy electrical loads as follows:

■ **If flaps down:**
Galley Power .... OFF
■ **If one generator inoperative:**
Pack..................ONE OFF
Select pack on same side as failed engine (if applicable).
■ **If two generator inoperative:**
Packs .... BOTH OFF
B Pumps ....ONE OFF

■ **If flaps up:**
■ **If one generator inoperative:**
Galley Power ....AS REQUIRED
■ **If two generators inoperative:**
Galley Power .... OFF

Restore: B Pumps ....ONE OFF
Load: Restore Bus..................AS REQUIRED
KW Loads ....CHECK

Hydraulic Fluid Quantity Decreasing   (15-9):

System A Hydraulic Quantity below 2.5 Gallons and Decreasing
Leak in System A
Both SYS A Pumps .... OFF
Both SYS A Fluid Shutoffs .... CLOSE

System A Hydraulic Quantity Stable at 2.5 Gallons and B Quantity Decreasing
Leak in System B
Both System B Pumps .... OFF

Maneuvers

This section is a consolidation of emergency maneuvers. Pilots are expected to be proficient in the performance of these maneuvers.

Editor's Note: *This study guide contains only selected maneuvers. See OM Maneuvers Section for complete details and notes. "Red Box" maneuvers are included in the red box section on page 10.*

Unusual Attitudes   *(MANEUVERS-7):* RECOVERY
AUTOPILOT .... DISCONNECT
See other notes on recognition, nose high and nose low recoveries in the MANEUVERS section, page 7.

Windshear/Microburst During Takeoff Roll *(MANEUVERS-5):*		■ **If the decision is made to continue the takeoff:** Throttles—Full Forward Increase V$_R$ by up to 20 knots Begin rotation no later than 2000 feet from the end of the runway even thought the adjusted V$_R$ has not been attained. ■ **If windshear conditions still evident once airborne:** Accomplish Windshear/Microburst Go-Around Procedure
Emergency Descent	*(MANEUVERS-8):*	CONTINUOUS IGNITION ....ON NO SMOKING and FASTEN SEAT BELTS SWITCHES ....ON THROTTLES ....CLOSE SPEED BRAKE.. FULL AFT AUTOPILOT ....OFF DESCEND....................STRAIGHT AHEAD or BANK (30º MAX) Ensure level-off altitude provides adequate terrain clearance TARGET AIRSPEED ....MAX .88M or VMO Unless structural integrity in question. TRANSPONDER.......................................CODE 7700 ATC ....CALL PASSENGER OXYGEN SWITCH (if required) ....ON/CK LT ILLUMINATED MAKE PA—Use oxygen (if required)—fasten seat belts—no smoking. When cabin altitude at or below 10,000 feet, advise Flight Attendants via PA the oxygen masks may be removed. See additional notes, page MANEUVERS-8
Stall Recovery	*(MANEUVERS-9):*	**Immediately and simultaneously—** AUTOPILOT...................................... DISCONNECT PITCH ATTIDUDE ....REDUCE If risk of ground contact, do not reduce pitch attitude more than necessary to allow airspeed to increase. THRUST...INCREASE ■ **If ground contact is imminent:** Advance the Throttles to the full forward position to obtain maximum thrust. WINGS ....LEVEL Return to assigned altitude and proper configuration speed. See additional notes, page MANEUVERS-9
Moderate to severe Turbulence Encounter *(MANEUVERS-9):*		**Immediately and simultaneously—** CONTINUOUS IGNITION ....ON AUTOPILOT ....DISCONNECT ELEVATOR See additional CAUTIONS (4) and NOTE (1), page MANEUVERS-9 SPEED ....280 kts OR .80 M See additional CAUTION (1) and NOTE (1), page MANEUVERS-9 YAW DAMPER ENGAGE SWITCH ....ON SEAT BELT SIGNS...ON MAKE PA— *"Flight Attendants, secure the service and be seated immediately."* Refer to section 3a—CONDITIONALS for additional information on Turbulent Air
TCAS RA	*(MANEUVERS-10)*	**Corrective RA (Any Climb or Descent Aural Alert)** **Immediately—** AUTOPILOT..OFF See additional notes, page MANEUVERS-10.

5 – Air Conditioning and Pressurization

General

Definitions
Air Cycle Machine
Compressor/Turbine Combination
Pack Includes:
2 Ram air exchangers
1 Air cycle machine (Turbo-Compressor Expansion cooler)
Pack fan
Associated Ducting

Flow sensing Venturi Controls Modulation Valve
Decreases 8th stage flow if too much
Adds 13th stage air if 8th stage not enough

Pack Fans
Ground--ON when Packs on
In Flight--Auto ON:
Flaps (Inboard) out of full up
Pack Switch ON

Overheats & Trips

Two Types of Trips, One Overheat
Pack Trip--Three Causes
Compressor Discharge Temperature
Reaches 200º
Some Aircraft--Cooling doors open automatically if CDT over 115ºC
Supply Duct Temperature reaches 250ºF
Turbine Inlet Temperature overheats
3 Pack Trips--2 Gages
Compressor discharge trip-Pack temperature gage
Supply duct trip-Duct Temp gage
Turbine inlet trip-no gage
Bleed Air Trip: Overheat of air coming into pack valve
Too much hot 13th stage air
Could be a Pre-cooler failure
Overheat, Not Bleed Trip on #2 Engine
Only 8th stage air (no 13th) taken, so not as hot
13th stage approx. 800ºF, 200-225psi
Duct Overheat: Supply Duct Air over 190ºF
Not a trip--nothing turns off;
Mix valves drive full cold, however

Corrective Actions
General--After trip:
Put switches in agreement with valve positions after trip
Hit "reset" to regain control of switches
Packs & Supply Duct: *Pack Up, Duct Down*
For a Pack Overheat (Trip),
Temperature up--Pack is working too hard
Should only reset a tripped pack once during a flight
For a Duct Overheat (NO Trip--just overheat light),
Temperature down to cool duct
Ducts Unlimited--No limit on number of resets

Bleed Trip:
Turn off Affected Bleed
Gets switch in agreement with the closed valve
Reconfigure bleeds to get air to packs, if necessary

Three Lines of Defense Against Supply Duct Overheat
Topping Circuit--Only in AUTO
140ºF in Supply Duct
Prevents any hotter position of the mixing valve
Temperature limiting value
190ºF in Supply Duct
Sensor near entrance to duct
Drives mixing valve full cold
Turns on *Duct Overheat* Light
Correction:
Select Cooler auto temp
Press reset
Absolute Temperature Limit
250ºF in Supply Duct
Turns off pack
Runs mix valve full cold
Turns on *Pack Trip Off* Light
Any time the **pack valve closes** for **any reason**, the **mix valve** goes **full cold**
Otherwise--first air coming into the supply duct after a pack is turned on would be full hot until pack started putting out cold air

Split Power Restrictions
Split power definition: Different sources for air and electric's, i. e.,
APU providing Air to Packs
External Power providing Electrical power
Requires cockpit monitor--flight crew or mechanic
Electrical power loss removes power from cooling fan, and pack could overheat

Auto Pack Trip
Armed when: (**GAFT**)
Ground-- *A/C on Ground*
Auto Pack Trip Switch--*NORMAL*
Flaps not up
Throttles all above 1.5 EPR
Has separate sensor--doesn't rely on EPR gages
Activated when any engine EPR drops below 1.3
Actions resulting from Auto Pack Trip
Turns Pack Fans OFF--Reduces electrical load
Closes Pack Valves--Reduces Engine Bleed load on remaining engines
Turns on Engine Failure lights
Following Auto-Pack Trip, if not reset:
Both Packs come on automatically and simultaneously at flaps out of Up
Could overload generators

APU Flow Multiplier Overheat Sensor

Turns on APU Bleed Light
Flow Multiplier & APU Bleed Valves Close
Maintenance check recommended for further use

Operating Considerations

Air Package Operation

Primary heat exchanger--uses ram air for first stage cooling of bleed air
Air Cycle machine (turbo-compressor)
    Compressor packs air in, and temperature rises
    Heat is dissipated as airflow gives up energy to turn the turbine which turns compressor
Secondary heat exchanger
    Takes compressor discharge air and cools it with ram air before it is further cooled by the turbine
Partially cooled (but still warm) air from the primary heat exchanger is mixed with cold air from air cycle machine to provide moderated temperature air for cabin heating and cooling

Preheat of Cold-Soaked A/C

For initial cooling, Auto position will drive supply duct temperature too hot trying to heat aircraft
Use Manual valve position & look for duct temp about 100°F
Reselect AUTO when leaving cabin (for preflight or cabin inspection, etc.)

Cabin Air Distribution Lever Problems w/Overhead Position

Most effective heating or cooling on ground
High power settings in flight—too much air at face level
Can put out moisture through overhead ducts when cooling

Aft Zone Temperature Switch

Four things needed for proper operation--(VASS)
    Valve--Air Mix Valve About 1/4 travel remaining toward the cold side (for cooler in rear only)
    Automatic on temperature rheostat
    Sidewall--Cabin Air distribution lever in sidewall
    Stabilized Flight (cruise)
For warming rear, first two above don't apply
Small movements are best
Aft Cabin Zone Temp Light
    Signals a sidewall duct overheat
    Drives forward or aft zone temperature control valve toward closed
    Goes out when sidewall duct cools
Aft Cabin Temperature Zone Override Valves Closed Switch
    Locates "center" closed position of zone temp valve position indicator
    Simulates an overheat (Overheat light illuminates
    Causes open valve to close when button held depressed for 30 seconds

Cross References

2-Lim-5	Limitations
3-Starting-16	If fumes, packs OFF on power back
3-Clb-Cruise-Desc-4	Cooling doors operation
3-Starting-6	Use of pre-conditioned air for parking
3-Parking-7	Cabin air distribution lever when parking
3A-Cond-15	Cold Soaked Aircraft
4-Emerg-15	Air Conditioning Smoke
4-Emerg-16	Smoke Removal
5-Air-23	Flow multiplier overheat (APU bleed lt.)
11-Fire-13	Lower Aft Body Overheat on Ground
20-Eng-2	Engine Fail warning lights

Pressurization

General

Packs provide pressurized air to Cabin and some lower body compartments
Pressurization is modulated and controlled by restricting outflow
Protective devices prevent cabin over pressure or negative pressure

Cabin Pressure Change--Maximum Rates Desired

Pneumatic System Rate Limit
    Climb cabin at 500 FPM
    Descend cabin at 300 FPM
Electronic System Rates
    Controlled based on PSI change per minute
    Rates in FPM may be faster than rates above

Ground Venturi

Fan creates negative pressure
Keeps outflow valves open on ground

Outflow Valve Functions -- (NARC)

Negative Pressure relief
    Set at -1.0 psi

Altitude limiter--Set at:
    13,000' $\pm$ 1500', Pneumatic System
    15,000', Electronic System
Relief of excess Pressure (Pressure relief)
    Occurs in 8.9-9.6 psi range
    Dumps to 8.0 psi rapidly
    Pneumatic System only—Electronic System uses Safety Relief Valves
Cabin pressure regulation
    See Chart, p. 5-Air-12

Key Altitudes & Pressures

10,000'--Warning Horn
    Note--*Also raises the PA Volume*
    Reset: Cabin Descent through 9,500'
--Max dispatch altitude with pressure problem
13,000 $\pm$ 1500'--Altitude Limiter, Pneumatic Pressurization; Outflow Valves Closed
14,000'
    --Emergency O_2 System Activation (pneumatic and electrical switches)
    --Electronic Pressurization switches from AUTOMATIC to STANDBY Mode
    --Max altitude after unplanned depressurization

15,000'--Altitude Limiter—Electronic Pressurization; Outflow Valve Closes

8.6 psi--Auto Controller Limit (+0.15)

9.6 psi--Outflow Valves open in a range from 8.9-9.6psi; dump down to 8.0psi

Ground Pressurization

250' Below actual altitude (□ 0.125 psi)

Allows door opening in emergency

Power Requirements--

By System

Pneumatic System

Pneumatic Air Pressure

Cabin Altitude Warning Horn--Battery Transfer Bus

Ground Venturi Fan--Non-ESS AC

Electronic System

Auto--   AC #1 & DC #1

Standby--   AC #2 & DC #2

Manual AC-- Essential AC

Manual DC-- Battery Transfer Bus

By Power Source

Battery power

Pack Fan Switches & Pack valves

Cabin Altitude Warning Horn

Manual DC Electronic Pressurization Outflow Valve control

ESS AC

Manual AC Electronic Pressurization Outflow Valve control

115V NE AC

Pack Fans

All other valves and switches

Ground Venturi Fan

Flight/Ground Switch

Flight--

Drives Outflow valve toward closed

Then it modulates automatically

Cycling off & on controls rate of initial pressurization "spike" on the ground & prevents exceeding maximum target rates

Ground-Opens outflow valve, depressurizes cabin

Cargo Heat Outflow Valve

Closing valve decreases air loss by Approx. 40%

Rate Knob Function

Controls speed of movement of outflow valve

Full Increase--causes valve to move as rapidly as possible to correct to what is set on the Cabin Pressure Controller

Full Decrease--Essentially freezes the outflow valve at its present position

Electronic Pressurization

Inputs to Automatic mode

Cabin Pressure

Cruise Altitude

Barometric Pressure

Gear Squat Switch

Auto Fail Light Activation--

Rate too high (Over 2000 FPM)

AC Power Loss

Goes out if restored within 15 Sec

If not restored by 15 sec, goes to STBY mode

Cabin Altitude 14,000'

Also triggers a switch to STBY mode

Off Schedule Descent Light

Activated by aircraft descent prior to reaching scheduled level off

Returns cabin pressure to Departure field altitude

Set actual altitude in FLT ALT window to reset

Remembers departure field pressure altitude until Cruise Altitude is Reached **OR**

FLT ALT window reset to present flight level

After set cruise altitude is reached, aircraft descent triggers cabin descent to destination field pressure

Standby position

On electronic system--works like manual knob on pneumatic system

Rate of climb 50-2000FPM

Rate knob only works in Standby mode

Manual Pressurization

Man AC-Moves valve full range in **4** sec.

Man DC-Moves valve full range in **8** sec.

Pressure Relief

Accomplished with two safety relief valves

Pneumatically activated

Limit differential pressure to 9.6 psi.

Cross References

2-Lim-4	Limitations
3-Taxi-T/O-10	Pack Cooling Door ops with snow/slush covered runways
3- Taxi-T/O-11,12	Packs-Off Takeoff
3-Clb-Cruise-Des-4	After takeoff, 500 FPM for faster cooling
3-Clb-Cruise-Des-4	Close cooling doors ½ at 18,000'; close fully at cruise
3-Clb-Cruise-Des-5	Rate knob full INC at cruise
4-Emerg-35	Cockpit window damage--max psi (Put note cross-referencing chart on p. 5-22)
5-Air-21	Static Check
5-Air-23	Cabin differential pressure for T/O
5-Air-24	Target rates of climb/descent

6 – Anti-Ice and Rain Protection

Icing Conditions

Ground--8°C or 46°F OAT and:

Visible moisture, to include

Clouds; Fog with 1 mile or less visibility

Rain, sleet, snow, etc.

Surface moisture to include Standing Water, Slush, etc.

Flight--10°C or 50°F TAT and visible moisture, as above

Pre-Heating on Ground

Engine Anti ice

Must be on at least 20 seconds before applying T/O power to ensure engine/bleed stability

If T/O not within 10 minutes of starting:

Run each engine up to a power setting as high as possible 75% N_2 minimum desired

Engines should be run up "momentarily," once every ten minutes in icing conditions prior to takeoff

If icing encountered after adding T/O power with engine icing *OFF* wait until higher of:

☐1000' AFL   ☐Obstacle clearance

Wing Anti-Ice

Never on for T/O

Lowest allowed altitude is 1000' AFL

Wing Anti-ice on for 1 minute after reaching 1000' AFL when taking off with:

Water or slush on runway and

Temperature below 35°F/2°C

Lower Aft Body Overheat Light

Sensing Loop in 3 Areas

☐Keel Beam   ☐Aft Airstair area

☐Aft Cargo Area

Problem--hard to identify source

Sources for Anti-Icing Air

Engine Anti-Ice

8th stages of #1/2/3   ☐Inlets

13th Stages of #1/3   ☐Cowls

6th & 13th Stages of #2   ☐S-Duct

Wing Anti-Ice

6th & 13th stages of #1/3--LE devices

Bleed for Packs

#1/3--8th & 13th stages

#2--8th stage only

Wing Anti-Ice

Don't use on takeoff below 1000' AGL

May be used for landing, but:

G/A EPRs must be computed for Wing A/I *On*

Don't use if OAT over 10°C

Reason: Can damage hail-resistance characteristics of leading edges

Exception: About to enter icing conditions

Engine Anti Ice RPM Requirements In Flight

When penetrating or operating in icing conditions: Minimum--**55% N_1**

Moderate to severe icing (TAT below -6.5°C/20°F)--**70% N_1** or higher, except as required for landing

Pitot Heat Meter

Light is ON w/switches off

Switch ON--Light indicates

Pitot heater for Capt or FO is not being heated (drawing power)

Rotary Pitot Static Heat Selector

Selects Amperage displayed on Pitot Static Amp Meters

Meters should indicate:

Static Heaters

Pitot Probes

Capt, FO Pitot Head

Amp Meters balanced

T(emp)-**Probe Aux P**(itot Tube)

Rosemont Probe (TAT Probe)

Imbalance, Left High

Elev P--Elevator Feel probes on vertical stab (Balanced)

Window Heat--4 switches

Windows L1, L2, R1, R2

Switch on either window 1 or 2 on respective side also turns on heat to windows 4 & 5 on that side

Cross References

3-General-10	Packs off when landing at fields with elevation over 2000' MSL when using anti-ice
3-Pre-Flight-15	Lavatory Leaks
3-Starting-7	Start all engines & keep running in icing conditions
3-Taxi-T/O-10	Pack cooling doors closed for T/O with Slush
3-App-Ldg-G/A-4	Packs off on approach after descent through icing conditions
3-After Ldg/Parkng-4	Leave flaps 25° after landing in snow/slush
3A-Cond-15	Cold weather procedures (15-21)
3A-Cond-19	No Pack TO Required for A/I fluid sprayed in intakes
4-Emerg-18	Strut Overheat
4-Emerg-18	Lower Aft Body Overheat
3A-Cond-17,19, 14-Fuel-15	Fuel Heat Requirements
5-Air-31	Pressurization problems with wing anti-ice On
20-Eng-13	Frozen Start Valve Procedures
MEL 30-12	Dispatch with window heat INOP

7 -- Autoflight

General
Attitude inputs to Autopilot
 #2 VG (Except Ex-Brannif, N716-731)
Heading source
 Manual mode--heading info from #2 DG
 Other modes--heading info from Captain's CDI
 #2 DG, with CDI Select in Normal or F/O CDI Alt
 #1 DG with Capt CDI Alt
 All other modes--same source as Capt's CDI azimuth card
Aileron channel must be engaged before elevator channel
Rudder switch turns off respective yaw damper

Interlocks--Aileron Channel
Prevent Engaging (*EMPTY*):
 Mode Selector **not** in **M**anual
 Turn Knob **not** centered
Disengage **OR** prevent Engagement
 Put **D**own **Y**our (*EMPTY*) **V**anilla **C**one
 Power--lost on No. 2 Radio Bus (ESS, on Ex-Brannif)
 Disconnect Button on Control Wheel pushed
 Yaw Dampers--Both Off
 VG--Loss of power or fast erect cycle entered
 CDI Select Switch moved out of Normal

Interlocks--Elevator Channel
Prevent Engaging *or* Disengage (**ACTS**):
 Aileron Channel off
 Cruise Trim
 Cutout switch to Cutout
 Cruise Stab Trim switch moved
 Trim Switch Moved (on control yoke)
 Servo Switched (A to B, etc.)

Aileron Actuator
Uses Aileron Trim motor

Pitch Actuator
Uses Cruise Trim motor to reposition stabilizer
Two elevator channels--using hydraulic servos

Disconnect button on stick
Disengages auto pilot
Extinguishes & resets Autopilot disengage lights

ILS Approaches
Coupler Uses left (#1) VHF receiver
Back-Front Switch (below RMDI) inputs to ILS coupler, but use prohibited on Non-precision approaches

Yaw Dampers
Two channels
Hydraulics--same as source to rudder
Power--
 Upper -- Non-Essential
 Lower -- Essential
Lights--
 Respective yaw damper switch off
 Electrical power lost to respective yaw damper
 Considered *auto pilot brain* lights--(Per some G/S Instructors)
 Lights don't come on for hydraulic problems
Autopilot won't work with both yaw dampers off
Yaw Dampers Don't counter rudder pedal inputs
Rudder Pedals don't move to reflect yaw damper inputs

ILS Autopilot Desensitization
Desensitization begins normally:
 ♦ Following G/S intercept
 ♦ At 1500' AGL on radio altimeter
Occurs on a timed basis
 Localizer desensitizatizes over 150-second period
 Glide slope
 Desensitizatizes over 100-second period
 Middle Marker signal retriggers further desensitization for the glide slope extend portion of the approach
 Glide slope intercept prevents testing of the radio altimeters
Caution: *With Middle Marker out of service, or not installed, for acoupled approaches desensitization schedule will not be properly triggered. Inappropriately large corrections can result.*
INOP Radio Altimeter
 Desensitization occurs at G/S intercept
 G/S intercept early (outside the outer marker)
 Will cause early desensitization
 Placing Mode Selector momebtarily to MANual mode at 1500 feet will retrigger proper desensitization
High intercept doesn't key early desensitization with operating Radio Altimeter

Some aircraft
No auto pilot test feature
(SYRI, not SYRIA--See Section 8 Notes)

Cross References
2-Lim-5   Limitations
3-App-Ldg-G/A-8   FE Monitoring/Callout Items

8 -- Auxiliary Power Unit

Automatic Control Features

APU Automatic Shutdowns (FOO)

*F*ire Loop activated
*O*ver speed
*O*il pressure low

APU Overload

Cuts back on fuel on start to control EGT
Becomes load control after APU on speed, using air flow to control load after start

Fuel

Unshrouded fuel line in wheel well
APU Fuel Solenoid Valve
    Can't take off w/valve pointing forward--open
    Red handle points back (away from tank) when Valve closed
    T/O warning horn sounds if APU fuel valve open and throttles advanced
    Held open w/battery power
Fire handle & controls -- L Wheel well
Red Disk forward of L Main Gear indicates thermal discharge of APU freon bottle

Tailpipe Thermocouple

Self-generating reading of EGT

Control Thermostat

Controls fuel *during start* to keep EGT in limits
Continuous run function
    Cuts in as "on speed" condition is sensed
    Fuel Flow Modulated to keep Speed constant
    Control Thermostat Modulates Bleed air to regulate EGT as speed is maintained

Starting

Starting Features

Automatic cycle
APU Crank light is battery operated and will be on for APU start
Ignition within 15 seconds after crank light
Oil pressure rise opens fuel valve
Stabilization within 30 sec. (Total of 45 seconds) after light off
    Indicated by EGT, not power
Max EGT 710°C

APU Switch

OFF
    APU Fuel Tank Shutoff Valve Closed
    APU Fuel Solenoid (at fuel control) Closed
    Bleed Air Valve Closed
    APU Generator Breaker Tripped, Generator Breaker Light deactivated

ON
    Four Items Above Reversed
    APU Light on F/E Annunciator Panel ON (Indicates Fuel Solenoid Valve Armed)
    Generator Breaker Light Activated
        ON if interlocks satisfied
        OFF when APU GB Closed (APU Powering Sync Bus

Start Relay Interlocks

To remember, relate to the **Bold items** on the APU Control Panel

Battery switch on--**Start Switch**
Both fire handles in the "IN" position **--Fire Handle**
Fuel valve open--**Generator Breaker Switch** (*Same shape as boost pump switch*)
Reset of auto-shutdown after last activation--**Fire Test Switch**
Stop switch (L Main Wheel Well) in the "RUN" position--**Guarded Auto Fire Shutdown Switch** (*Looks like guarded switch in wheel well*)
Wheel well duct continuity loop good--**Circle of EGT Gage** *Looks like a Loop*

APU Start Sequence (p. 8-APU-9)

Start switch to START
Battery Charger Disconnected from Battery
    Note: *Monitor Battery voltage above* <u>22 Volts</u> *before start. If Below 16 Volts during start, watch for hot start.*
Starter engaged, CRANK light on
Oil Pressure rises, triggering:
    Ignition
    Fuel Solenoid Valve at fuel control OPENED
Lightoff Occurs, RPM & EGT rise
    Should occur within <u>15 seconds</u> *of CRANK light*
Starter disengages at dropout speed, CRANK light goes out
Slightly below governed speed:
    Ignition de-energized
    Bleed Valve Opened if at least one #2 Engine bleed valve is open
    Generator Breaker can be closed once Volts & Freqs are verified stable
    Should occur within <u>30 seconds</u> *of EGT rise*
APU Cooling after failed start attempt
    2 Minutes after aborted hot start
    5 Minutes after 2nd attempt
    3 Tries only allowed before maintenance required

Normal Operations

Only Run *from* 5 minutes before engine start *until* after all three engines started for takeoff (Capt may taxi on one or two engines and start remaining engines before takeoff)
Shut Down after:
    Last Engine to be started is running
    2 Minutes' cooling time after bleed air taken from APU
Guide--Look for APU EGT ≤350°C (Approx.)

Electrics

APU Generator Breaker Automatic Trips (APU'S Field)

A*lternate source powering Sync Bus*
P*hase Unbalance with APU connected to Sync Bus*
S*peed Faults*
APU **Field** *Relay Tripped*

APU Generator Field Relay Automatic Trips

Feeder Faults
Fire Handle (APU) Pulled

Generator Breaker

Light tells condition of interlocks pre-start
Light tells condition of generator post-start
Selectable Generator display--APU or Ext Power load

Power Requirements

Battery
    Start Control
    Fire Detection
    Fire Protection
Powering Essential using the APU generator
    #3 A/C Bus Tie Breaker must be closed
    Same requirement for External Power

Power output

Favors Electrical
    Speed is modulated to regulate constant 400 ± 10 Hz frequency
    Bleed airflow is whatever amount that speed produces
Electrical
    Same type generator as main electric AC generators
    Other 3 Main AC Generators:
        Limited to 36 KW continuous for cooling
        Max instantaneous is 54 KW (for 5-minute limit)
    APU Limiting Amperage is 165
        Compares to approximately 57 KW
        Higher generated power allowed due to better cooling provided for APU generator

Fire Protection

Fire Detection

Sensor in APU Shroud triggers 4 warnings:
    Bell in cockpit
    Intermittent Horn in nose wheel well
    Steady Light on FE panel
    Flashing Light in Wheel well
Sensor activates 3 protective measures
    Closes APU fuel shutoff
Bell Cutout Button
    Terminates Cockpit bell and NW horn
    Causes wheel well light to go steady

Fire Test After Start (First Flight of the Day Item)

Fire warning closes fuel valve
Fire handle sends backup signal to same valve
Disarm system to test, or it will shut down APU
Horn/Bell must sound within 60 seconds of test initiation
If system INOP, APU may be started if a fire guard is in place

Pulling APU Fire Handle (BIG & FAT)

Last three are immediate; first three take a few seconds
B*LEED VALVE--APU Bleed Valve closes as airflow ceases*
I*SOLATE--Isolates APU in shroud*
    Spring-loaded Cooling air discharge valve closes as airflow decreases
G*EN--APU Generator falls off line as APU spins down*
F*UEL--Closes APU fuel shutoff valve*
A*RM--Arms APU Freon bottle*
T*RIP--Trips Generator Field*

Bleed Air

Bleed Air Power Requirements

AC non-essential power to:
    Bleed Valve
    Switches
    Bleed air pressure gages
    L & R duct isolation valves
    L & R APU isolation valves
Valve won't open until APU up to speed

Cross References

2-Lim-5	Limitations
4-Emerg-6,11	Fire
5-Air-5	Bleed Air Distribution schematic
8-APU-9	Min #2 Fuel Tank Quantity for APU operation
8-APU-10	Cool Down 2 Minutes
10-Elec-14	Electrical Control
11-Fire-1,7	Fire Handle operation
11-Fire-2	APU Bleed Light (Flow Multiplier Overheat)

9 -- Communications

Radio Assignments
VHF 1--Company/ACARS
VHF 2--ATC
SELCAL--
    Selective Call
    Allows monitor of all radios, while company calls for a "call-in" by SELCAL

Power Source Rule of Thumb
If only one of an item, on ESS
If two of an item--
    Captain's or #1 ESS
    F/O's or #2--Radio Bus
Ex Braniff A/C follow the rule to the letter
Exceptions--Pure AA A/C
    Autopilot--Radio Bus
    CADC
        #2 CADC on ESS
        #1 CADC on Radio Bus 2

Equipment Cooling
Fan draws air through E & E compartments
"No Equipment Cooling" light
    Indicates insufficient cooling air
Equipment cooling
    Fan runs until 0.5psi cabin differential pressure--enough to ensure airflow
    Equipment Cooling Valve (ECV) stays open until cabin differential press reaches 2.8psi, then closed
    Above 2.8psi, enough airflow through fixed bleed with ECV closed

Transponder
Alt reporting switch
Green light on for test or when interrogated by ATC

PA System
Light on panel indicates system in use by F/As
Capt PA announcements do not light the light

Flight/Cabin Interphone Switch
Permits communication between cockpit and cabin using
    Hand-held microphones *or*
    Oxygen mask mics
Two switch positions
    Normal—Flight interphone system operates normally
    Link—Allows communications to cabin station handsets using sources above
Switch location
    Overhead panel
    Below Flight Attendant Call Panel

NORMAL   LINK
FLT/CAB INTERPHONE LINK

10 -- Electrical System

Overview

General
Needed to generate power
  Magnet + Conductor
  Relative motion--Can be rotation or side-side
Field strength can be used to control generated power
KVAR
  KiloVolt-Ampere Reactive
  Measures reactive load
  "Cost of doing work"
Short Circuit
  Electricity going someplace you don't want it
  Amps=Volts/Resistance
  As Resistance goes to 0, Amps go max

B-727 Generator
<u>Small permanent magnet</u> Generates small Residual Voltage (10-17 volts)
  Residual Voltage Button shows residual voltage
    Must be read with generator field relay tripped, or actual generated voltage may burn out the gage
    Don't touch unless procedure requires (CSD Disconnect)
<u>Voltage regulator</u> senses 10-17, wants 115$\pm$5 volts
  V/R increases power at exciter to result in increased power output
<u>3-phase, 3-coil</u>
  120 degrees out of phase
<u>Automatic paralleling</u> of generators
  Occurs with Sync Bus Tie relay closed for respective generator
  Accomplished with Generator Breaker

Battery
Located in E & E panel
Presence/security may be checked on preflight if desired

AC Power

Generator Control Circuits
One Set for Each generator
Powered by Battery Bus when:
  Generator not operating
  Battery switch on
Powered by respective generator when generator operating, field relay closed

Synchronizing Bus (Sync Bus)
Compares voltages on different phases
Provides redundancy to Load Busses
Provides load-sharing

Control relays
<u>Bus Tie Breaker</u> for each generator (BTB) Separates respective AC bus from sync bus
<u>Generator Breaker</u> connects Generator to load bus
<u>Field Relay</u> shuts down generator by cutting voltage at generator field

6 AC Busses
#1 Main AC Bus
#2 Main AC Bus
#3 Main AC Bus
Essential AC Bus--Can only be powered from Bus 3 if using <u>APU</u> (APU Generator Breaker Closed) or <u>External Power</u> to power the Sync Bus
Standby AC Bus
  Normal--power from Essential bus
  "STBY" on Essential Power Selector--Battery bus through static inverter
AC Transfer Bus--Normally from #3 AC Bus
  Battery Charging
  Used for Ground Servicing

Essential Power
Any operating generator can be selected to supply power to Essential Bus
Power Taken from between field relay and generator breaker
Normally--#3 Generator is in-flight source

Master Warn Light On when:
Essential AC Bus power is removed (having been powered once)
Non-Essential AC power available but Essential AC bus is unpowered

Standby Bus--(#1^4 + CP)
#1 *VG (HDI)*
#1 *DG (RMDI)*
#1 *Comm*
#1 *Nav*
C*apt's CDI Alternate Source selection*
P*neumatic Valves (Flow Multiplier Bypass Valve)*

Sync Bus Tie Automatic Trips--EXCITER Problems (SEEP)
S*tability Protection*
E*xcitation--Over or under excitation*
E*xciter Ceiling Protection (not cleared by other protective circuits)*
P*hase imbalance will trip all three (Not Exciter-related)*

Generator Breaker Automatic Trips--SPEED Problems (FACED)
F*ield Relay of Generator Tripped*
A*PU GB Closed*
C*SD over speed or underspeed*
E*xternal Power on*
D*isconnect of CSD*

Generator Field Relay Automatic Trips--VOLTAGE Problems (VDF SEE)

Voltage--*Over or under voltage situation*
Differential Fault--*Feeder shorted or grounded*
Fire Handle Pulled
SEE--*From BTB,* **SEE** *items will trip Field if problem persists after BTB opened*

Key to Remembering Disconnects--(ESV)

Like ESP but electrical: last letter is V for Voltage

Sync Bus .... **E**xciter
Generator .... **S**peed
Field Relay... **V**oltage

APU Generator Breaker Automatic Trips (APU'S Field)

Alternate source powering Sync Bus
Phase **U**nbalance with APU connected to sync bus
Speed Faults
APU **Field** Relay Tripped

APU Generator Field Automatic Trips

Feeder Faults
APU Fire Handle Pulled

Ground Interconnect Electrical Switch

Items with a Ground-Only Function (**SYRIA**)
In Effect with either of the following:
  APU powering Sync Bus
  External Power *connected* (not necessarily powering busses
Stall Warning -- Deactivated
Yaw Dampers -- Deactivated
Recorder (Flight)--Deactivated
Interconnect (Ground)
  Activated--Hydraulic Ground Interconnect
  Needed on ground for hydraulic tests
Autopilot--Deactivated
Also--Fueling Valves Operational

White Circuit Breakers (VET)

Voice Recorder CB
Essential Bus Tie CB
Trim
  Main Electric (Fast) Trim CB
  Cruise & Autopilot (Slow) Trim CB

CSD

Frequency Knob

Only works when respective generator is not in parallel operation
Can vary cycles approx. $\pm$ 7, range of 15

Disconnect Switch

Requires engine above Idle to assure disconnect
Reconnecting
  Can only be reset on ground by maintenance
  Engine shut down required

Temperature Gage

In: Indicates efficiency of CSD Cooler
  Temperature of fluid coming into oil pump is read
  Critical at low altitude and high ambient temperatures
Rise: Indication of workload
  Difference between:
  Fluid temperature leaving cooler (entering pump)
  Fluid temperature leaving the pump (entering cooler)

CSD Oil Cooler

Old A/C--ejector valve actuated
  Ejector valve using 13th stage air draws ambient air through cooler
  "Ground Off"
    Ground--Normal is on, alt "GND OFF"
    Allows manual turn off of ejector air
    Preserves bleed
  "Normal"
    On for ground operations
    Off when gear squat switch indicates airborne
  "Flight On"
    Flight--Norm is off, alt "FLT ON"
    Used for slow speed flight
Newer engines--fan air cooling
  Oil stays 10-15° hotter than on older A/C
  Ground Off Position is deactivated, but same procedure is used for Standardization
  Can't tell from cockpit if engines modified
If Oil in "Rise caution range"
  Trip affected generator
  Monitor for 3 minutes
  If still in caution range, CSD -- disconnect
  If cools below caution range, Generator kept tripped and used as backup

High ambient temperature Operations

IN temp may be in caution range until cruise level

DC Power

Main Uses of DC/Battery Power

Indicating Lights
Radios
Part of Autopilot
Switches
Fuel Shutoff Valves (Battery)
Pneumatic & AC Valves (Battery)
Part of Ignition System (Battery)

4 DC Busses

#1 DC Bus--From #1 TR, which is powered by #1 AC Main Bus
#2 DC Bus--From #2 TR, which is powered by #2 AC Main Bus
Essential DC--From Essential AC Bus, powered by the selected AC source for ESS AC, and backed up by TR #1 & TR #2 if required
Standby DC-From DC Essential bus

Indicators for DC System

Amp meter--reads load on the TR
Volt meter--reads voltage at the Bus

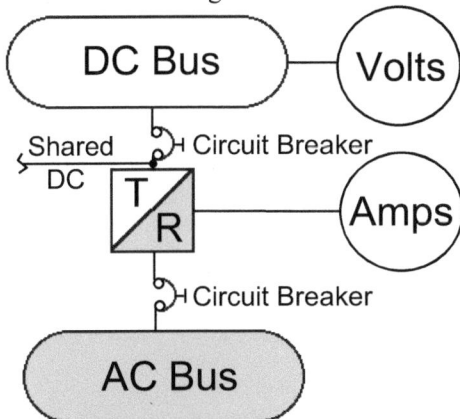

Therefore, if upper CB Trips (P6-A/I-2), Some amps from shared DC load, but zero volts on bus

Reverse Current Relays

Internal to TR
Protects each TR from over-voltage due to problems from the other TR

Blocking Diode

Protects essential bus from powering other two in case of failure
By definition, diodes block, so a redundant name
(A non-blocking diode is a wire!)
Allows either TR 1 or 2 to power essential DC

Standby DC Bus

Normal--Power from Essential DC Bus
Essential Power Selector STBY--Static Inverter powered from Battery Bus

Battery

5 Battery Busses

Hot Battery
Battery (ESS DC)
Hot Battery Transfer (ESS DC)
Battery Transfer (ESS DC)
Generator Control

Hot Battery Bus-Driven Busses

Switch Off-
  Hot Battery Bus
  Hot Battery Transfer Bus
Switch On
  Hot Battery Bus
  Hot Battery Transfer Bus
  Battery Bus
  Battery Transfer Bus
  Generator Control Bus
Essential DC Bus-Powered Busses
  Battery Bus
  Battery Transfer Bus
  Hot Battery Transfer Bus
  Standby DC Bus
Drains on Battery during Normal Ops
  Generator Control DC Bus (in standby status)
  Hot Battery Bus
  Therefore--little actual drain
Drains on Battery with Battery Switch OFF--Slight drain if voltmeter left in "Battery" position

Hot Battery Bus--(BAR BBGH)

Battery Bus
APU Cranking & Starting
Relay Control
Battery voltmeter & Switch
Bus Protection Panel
Generator control Panel
Hot Battery Transfer bus

Lighting On with only Battery Power

Four W's
Walnut Lights--Emergency Background Panel lights
Whiskey Compass
Walkout--
  Emergency Exit Lights over Door Exits
  Aisle Escape Path Lights (Floor Strip Lighting)
  Over-Wing Escape Slides, some aircraft
White Dome Lights

Cross References--AC & DC Power

2-Lim-5	Limitations
3-Starting-13,14	Electrical power buildup during normal start (See also 3-Starting-15)
3-Starting-15	Essential Power Priority
4-Emerg-8	Engine Fire (Red Box)
4-Emerg-10	Precautionary Engine Shutdown
4-Emerg-12	Essential Power drain is Approx. 6 KW
4-Emerg-12,13	Electrical Fire & Smoke
4-Emerg-16	Loss of all Generators
MEL 24-1	Dispatch with 2 Generators

11 -- Fire Protection

Note: *Much of the material in this section is also found in Sections 8 (Electrical) and 16 (Engines). It is consolidated here as well for ease of study.*

Engine Fire Warning & Protection

Fire Warning--General
Loops on all three engines & fire walls
Loops in wheel wells (1 light only)
Test only checks circuit continuity--does not heat circuits

Engine Fire System Electrical Power
Fire *Detection*--Essential
    Except APU Fire detection is Battery
Fire *Protection*--Battery
    Except Fire discharge **Bottle Lights**—Non-Ess. DC

Freon bottles
Button
    Arms, Aims freon toward engine w/handle out
    Discharges bottle when a fire handle is pulled
3 Disks on Right side under pod
    2 Red disks--thermal overpressure discharge
    Yellow disk either bottle discharged intentionally
Pressures--see limits

Power Sources
Hot battery bus
    Fuel shutoff switches..Fluid
Battery bus ....⇩
    Engine hydraulic fluid shutoffs.............................⇩
    Generator Field Relay ....Elect
    Freon bottle arming/Activation.............................⇩
Non-Essential AC
    Bleed Valves ....Air
    Engine anti-ice ....⇩
Essential AC: Fire detection system

Pulling Fire Handle:
Isolates engine from fuel, hydraulics, and unnecessary bleed
    air drain
2 Fluids
    Fuel cut at engine fuel shutoff valve
    Hydraulics: **A** System pump fluid shut off (exc. #3 eng)
        Disarms low hydraulic system panel pressure light
2 Electrics
    Generator field Opens after 5-10 sec.
    Arms Freon Bottle Discharge Switch
2 Airs
    Bleed air shutoff
        Same effect as turning off respective engine bleed
          switch(es) on A/C & pressurization panel
        Closes Duct Isolation Valve
    Anti-ice external to engine
        Wing Anti-Ice Valves Closed (1 & 3 only)
        Closes #2 Inlet Cowl Duct Anti-Icing
        Does not effect Engine anti-ice for respective engine,
          or for #2 Inlet
        Cowl (pod engines) & compressor inlet heat (all 3
          engines) stays on

APU Fire Warning

General
Fire test after start
Fire warning closes fuel valve
Fire handle sends backup signal to same valve

APU Fire Automatic Actions
Sensor in APU Shroud triggers 4 warnings:
    Bell in cockpit
    Intermittent Horn in nose wheel well
    Steady light on FE panel
    Flashing light in L Main Wheel well
Sensor activates 3 protective measures
    Closes APU Fuel Shutoff Valve
    Audible Warnings in Cockpit & NW Well
    Lights in Cockpit and L Main Well
Bell cutout
    Terminates NW horn also
    Changes flashing Wheel Well light to steady

Pulling APU Fire Handle (BIG FAT)
Last 3 immediate; first 3 several sec.
BLEED VALVE--APU Bleed Valve closes as airflow ceases
ISOLATE--Isolates APU in shroud; Spring-loaded Cooling air
    discharge valve closes as airflow decreases
GEN--APU Generator off line as APU spins down
FUEL--Closes APU fuel shutoff valve
ARM--Arms APU Freon bottle
TRIP--Trips Generator Field

Cargo Smoke & Fire Suppression System

Detectors--FWD and AFT cargo compartments
Uses photo-electric cells and particle detectors
Several zones in each compartment
Loops A and B for each zone

Alerting System—Activates when:
Both loops (A and B) in a zone detect fire or smoke
One detector detects fire or smoke and the other has a fault
Smoke detection occurs within 60 seconds of occurrence

Suppression system—For fire in either compartment
Two fire bottles—aft part of forward cargo compartment
Halon directed to affected compartment with diverters
Discharge occurs by pressing associated **FIRE** switch/light
BTL 1 discharges at high rate immediately
BTL 2 discharges 15 minutes later at slow rate

Power Sources
Loop A—28V DC Essential   Loop B—Battery XFer

Cross References
2-Lim-6 .... Limitations 4-Emerg-6..... Fire Emergency Chklist
4-Emerg-11 . APU Fire   4-Emerg-17 ... Wheel Well Fire
4-Emerg-18 . Strut & Lower Aft Body Overheat
5-Air-11 ...... -200A & -200B Do not have #2 Bleed Valve;
          Pulling fire handle closes both #2 duct isolation
          valves

12 -- Flight Controls

Trim Tab Types & Uses

Balance Tab--Small tab which moves in direction opposite flight control deflection

Control Tab--Makes control easier without hydraulic power--"Flies" control to new angle

Ailerons and Elevator have balance tab

Anti Balance Tab--Moves in same direction as flight control

Anti-balance tabs make control more effective, but manual operation more difficult

Rudder has Anti-balance tab

No problem, since no manual rudder control

Ailerons

General

A & B Hydraulic Systems Power ailerons

Systems back each other up in case of failure

Hydraulic Power Lost--Aileron Manual Reversion

Cable Unlock following loss of hydraulic pressure Allows manual reversion

Inboard Ailerons--

Tabs on inboard ailerons become control tabs

Cables Move control tabs

Aerodynamic forces on control tabs are then used to reposition the complete aileron system

Outboard Ailerons--Mechanical linkage to inboard aileron moves outboard aileron

1/3 More control yoke movement required to get the same aileron movement if unlocked

No aileron trim available with manual reversion

Spoilers

Spoilers operate to supplement aileron control within blow down limits

Spoiler Blow down

Begins at 250 KtsFull Spoiler Deployment available below this speed--45° deflection

Complete at 400 KtsOnly 45% of Spoiler Deployment available above this speed--20° deflection

Ground Spoilers

Inner two spoilers on each wing

Signaled by

Mechanical Z-Link on left main gear strut

Speed Brake handle out of full forward

Spoilers operate as speed brakes (except ground spoilers) when speed brake handle is pulled

No Manual Reversion for spoilers

Outboard Aileron Lockout

0° Flaps--0% outboard aileron

5° Flaps--80% outboard aileron

35° Flaps--100% outboard aileron

Elevator

General

Left & Right Elevators mechanically independent

Actuators are mechanically linked, therefore automatically compensate for loss of input to one actuator

Powered by both A & B systems

Powered--Hydraulics actuate elevator and balance tab

Unpowered Mode--Manual Reversion

Balance tab becomes control tab

Manual input moves tab

Full yoke travel results in only 60% travel of elevators

Elevator Feel

Feel computer gets inputs from Pitot Feel probes

Still works with loss of Sys A or Sys B, (But not loss of both)

Centering spring provides feel with loss of both systems

Differential Pressure Light indicates metered output different -- May be caused by:

Icing of one Pitot probe or

Loss of Hydraulic A or B at feel unit

Pitch Trim

Moveable Horizontal Stabilizer is repositioned through an electric motor or manually driven jackscrew

Not "elevator trim" in that neither elevator movement nor neutral point are moved

1 Unit □11 turns of trim wheel

Markings on Vertical Stabilizer

Nose down limit is upper mark on tail (-2.5 units)

0 Units--leading edge of tail plane at middle mark

Full Up trim (12.5 units)--leading edge of tail plane at lower mark

Neutral Shift

Shifts neutral position of yoke

Allows more elevator travel nose down as yoke is trimmed extremely nose up

Assists in slow-speed maneuvers such as go-around

Begins at 6 units nose up

Complete at 10 units nose up

Autopilot Pitch Inputs

Fed to one of two elevator actuators, depending on whether auto pilot servo A or B is selected

Mechanical mirroring linkage transmits commanded input to other side (See above)

Rudder

Yaw Dampers
Airplane Yaw sensed and compensated through commanded rudder inputs
More info in Autopilot Section

Power
Hydraulic actuator moves Rudder
Anti-Balance tab goes in same direction (*anti*-balance tab) as rudder
Improves aerodynamic efficiency of the rudder
Upper rudder powered by System B
Lower rudder powered by System A; Backup Standby System
No Manual Reversion!

Rudder Load Limiter Light
Indicates improper lower rudder pressure for flap setting

Low[4] -- With Flaps
Lowered, light indicates
Lower pressure to the
Lower rudder, and the X-Wind limitation is
Lowered to 10 knots
With flaps up, light indicates
Higher than normal pressure to the rudder
Avoid large movements of rudders

Backup to Hydraulic Sys A & B
Standby System powers Lower Rudder
Standby selection is essential power
No manual reversion for upper or lower rudder
Yaw Dampers INOP with A or B system loss

High Lift Devices (Flaps & Slats)

General
Triple slotted flaps
Flap position read from middle flap

Inboard Flap Switches (PARIS)
Pack Fans
Auto Pack Trip
Rudder Load Limiter
In flight Warning of Speed Brake+Flaps
Stall Warning

Outboard Flap Switches--GOLF
GPWS
Outboard Aileron Lockout
LED Sequence Valve
Flap+Gear Warning Horn

Jammed Flaps--
One set (inboard or outboard) stops at a setting short of that commanded. i. e. one set at 5°, other at 15° with flap handle at 15°
No isolation of hydraulic pressure
Must use alternate flaps to electrically lower other (unjammed) set

Asymmetric (Split) Flaps--
Needle split between L & R needle within a system (inboard or outboard)
Hydraulics automatically shut off from split set (inboard or outboard, not left/right)
Normal lowering of unjammed (symmetrical) set may be continued with flap handle

Alternate Method of Lowering Flaps
AC Motor Drives shaft connected to same Torque Tube as hydraulic pump
Guarded Master Switch
Arms Inboard & Outboard switches
Cuts off System A pressure to all flap actuating devices
Turns on Standby pump
Inboard & Outboard switches
Open shutoff to hydraulic-driven pump for LED
Trailing Edge Flaps have electrical backup (takes up to 6 times as long)
Limit--15° Trailing Edge Flaps if System A or B fails
Spoilers not held down--May float up & create drag if flaps are extended beyond 15°

30° & 40° Flaps Blocked--Some Aircraft
-200A—Removable Block to prevent inadvertant selection of flaps 30°
The block is hinged and safetied

Slats
Power
Primary--A System
Backup--Through Standby System
Flaps 2°--Slats 2,3,6,7 out
Flaps 5°--All slats out

Warning Horns
Most warning horns are flight control-related. They are found in various sections of the Flight Manual, and have been consolidated here for ease of study.

Throttle-Activated Takeoff Warning Horn (FLASS)
Warning horn sounds if any of the following are not in the specified position:
Flaps--Takeoff setting of 5°, 15° or 25°
Leading Edge Devices--Down, Green light ON
APU--Off
Speed Brake Handle Forward
Stab Trim in the Green Range

Gear--Always a STEADY Horn if Gear-Related
Flaps over 27°* with any gear not down & locked
Cannot be silenced without gear down or flaps repositioned to less than 27°*
Gear not down with throttles retarded near idle
Silenced with "Horn Cutout Pull" lever
*Railsbeck Kits installed—22

Cabin Altitude Warning--
For Cabin over 10,000' pressure
See Section 5 Notes

Speed Brake Warning Horn

Warns of simultaneous deployment of flaps and speed brakes
in flight

Speed Brake Warning Horn Sounds when:

Speed Brake Handle not fully stowed

And Either of the following:

Flaps not full up

Throttles advanced more than 2/3 on the ground

Cross References

2-Lim-6	Limitations
3-Clb-Cruise-Des-9	Wind & Gust Additives
3-After Ldg/Parkng-2	Stab Trim Set at 5° or 0°
3-Taxi-T/O-1	Flap Operations with snow, slush or ice (See also 3-After Ldg-Parking-4)
4-Emerg-4	Voice Recorder CB Pulled for Flight Control Malfunctions
4-Emerg-22	20° Maximum Speed Brakes use with 1 Main and Nose Gear Extended
4-Emerg-47	Jammed Stabilizer LED Asymmetry or Jammed Flaps
4-Emerg-48	Wing Flap Asymmetry or Jammed Flaps
4-Emerg-49	Manual Reversion Landing
17-Misc-2	Warning Horns

13 -- Flight Instruments

Primary Instruments
Pitot Static Instruments
Compass
Attitude Indicators

Secondary Flight Instruments
Comparator
Radio Altimeter

Basic Corrective Actions
"Boostrap" instrument to another source
Select alternate source of power
Fly home with what's left

Instrument Information Sources

CADC (Central Air Data Computer)
#1 CADC Feeds:
- ◆ Capt Alt, A/S
- ◆ GPWS
- ◆ Autopilot
- ◆ ATC Transponder #1

#2 CADC Feeds:
- ◆ F/O Alt, A/S
- ◆ Flight Recorder
- ◆ ATC Transponder #2

Pneumatic System Directly Feeds:
- ◆ IVSIs
- ◆ #3 Alt, #3 IAS

VG (Vertical Gyro--Attitude Information)
HDI (Horizon Director Indicator)
#1 VG
- ◆ Capt HDI
- ◆ Radar, RDR-1E (No Bootstrap)
- ◆ Capt Horizon Amplifier (Bootstrap capability)
- ◆ Capt Steer Computer (No Bootstrap)

#2 VG
- ◆ F/O HDI
- ◆ Radar, RDR-1F (No Bootstrap)
- ◆ F/O Horizon Amplifier (Bootstrap capability)
- ◆ F/O Steer Computer (No Bootstrap)

DG (Directional Gyro; Heading Information)
RMDI=Radio Magnetic Deviation Indicator
CDI=Course Deviation Indicator
#1 DG--Capt RMDI, Slaved==>F/O CDI
#2 DG--F/O RMDI, Slaved==>Capt CDI

Note: *This information is found in the NAV section. Placement here is for ease of study with related systems.*

FD-108 (Flight Computer)
Off--V-Bars out of view
HDG--V-Bars Command bank angle to steer to the CDI cursor-designated heading
Loc--Heading commands only & no bootstrap
G/S--ILS pitch & bank commands to V-bars
  Amber Loc light when selected
  Green Loc light when locked
  Amber G/S light when Loc light goes to green
  Green G/S light when G/S intercepted
HDI "Computer" flag refers to FD-108 failure in selected mode or power failure any time

Radio Altimeter
Most Aircraft:
  Only one radio transmitter
  Test one, and other reacts as well
Indicator failure doesn't stop signals to GPWS
Key Altitudes
  18,000'--Turn on, Test, & Set for approach
  2,500'--Tape appears
  1,500'--GS is captured; Desensitization occurs
  1,000'--DH light on; push to extinguish
  500'--Arms Go-Around Mode
  DH--Light comes on at decision height set and at 1000' AGL

Power Loss

Rule of Thumb for Power Loss
If there is only one of an item, power is Essential
Two of an item:
  Capt on ESS or STBY
  F/O on Radio Bus #2 (Non-ESS)
Exceptions:
  Single items not on ESS:
    Radar on Non-ESS Bus
    Autopilot not on ESS
  CADC #1 on Radio Bus #2
  CADC #2 on ESS
    (Except N716AA - N731AA, #2 radio bus)
  Ex Braniff don't have the above exceptions
  ADF #1 on Extended Over waterESS
  ADF #2 on Domestic models--ESS
  Note: *For other information on power losses, see Section 14 Notes.*

GPWS (Ground Proximity Warning System)

General
Includes both Visual & Aural Warning
GPWS Monitors:
  Descent rate when below 2500' AGL
  Closure rate with terrain
  Flight path during T/O & initial climb
  Gear/flap position below 500' AGL
  Position with respect to glide slope
No Warning Of:
  Short landings if on a normal descent rate
  Sharp Terrain features such as cliffs
Two Warnings
  First hard"Whoop-Whoop Pull Up"
  Second softer--"Glide Slope"--(no "Whoops")

Computer inputs
- ◆ Radio Altimeter inputs from #1 radio altimeter, regardless of switch position
- ◆ #1 CADC
- ◆ Flap & Gear
- ◆ #1 ILS Receiver
- ◆ Stall warning inhibits

Rejecting warning (*disabling system*) authorized for:
Emergency Gear Up landing
Ditching
Non-standard flap configuration landing

GPWS Light
Self test detects a failure
Warns of invalid Input signal
Computer fail makes light come on
Do not disable unless Ops Manual Procedure specifies that course of action

Pull Up light (Red)
Comes on flashing for mode 1 through 4
Accompanied by aural warning
To be Replaced by GPWS light (PB-215)

Below Glide Slope (Amber)
Steady for Mode 5 Alert
Accompanied by aural warning

Windshear INOP (Amber)
F/E Annunciator Panel
Indicates windshear computer failure detected

Windshear (WINDSHR) Alert Light
Decreasing performance--Upper half ON--red
Increasing performance--lower half ON--amber

System Test
Tests light and warning tones
Can't repeat test for 10 seconds

7 Modes
Mode 1--Excessive Descent Rate
Excessive descent rate below 2500'
Roughly 3 to 4 times altitude; i. e. 1000' AGL occurs at ≈ 3500fpm sink rate
No configuration limits
Timing based on 30 seconds to ground impact
Loosened criteria below 1180' AGL to prevent nuisance alerts
Inhibited below 50 feet
Aural Warning--"SINK RATE" or "Whoop Whoop PULL UP" depending on rate
Mode 2--Excessive Terrain Closure Rate
Excessive terrain closure based on Radio Altimeter
Mode 2A--Less sensitive--for approach phase
Flaps not in landing configuration.
Not on glideslope ±2 dots of glideslope
Mode 2B--Same as above, but flaps landing configuration and within ±2 dots of glideslope
If gear up or flaps not in landing configuration, "Terrain" & "Whoop-Whoop Pull Up" warnings sound
Mode 3--Altitude Loss After T/O or Go Around
Descent during takeoff regime prior to reaching 2450' AGL if 10% altitude loss occurs
Activates on go-around from below 245' AGL
Aural Warning--"DON'T SINK"

Mode 4--Unsafe Terrain Clearance
Not in landing configuration below 500' AGL (fighter pilot hell)
Three sub-modes
4A--Terrain Proximity, Gear Up
4B--Terrain Proximity, Flaps Up
4C--Minimum Terrain Clearance, Gear & Flaps not in landing configuration
Aural Warnings for each mode--"Too Low Terrain," or "Too Low Gear" depending on parameters encountered
Mode 5--Deviation Below Glide Slope
Valid glide slope signal required
Below 1000 AGL
Below 1 1/3 dots low--"Glideslope" soft alert
Below 2 dots low--"Glideslope" hard (louder) alert
Mode 6--Altitude/Bank Callouts below 100' AGL
Altitude Aural Warnings
Replace pilot-required 50, 40, 30, 20, & 10 Foot callouts
Bank Aural Warnings
Above 300' AGL--Occurs between 31 & 40° depending on roll rate
Below 300' AGL--progressively lower bank angles; depending on roll rate & altitude
Aural Warning--"Bank Angle, Bank Angle"
Mode 7--Windshear Detection
A/C below 1500' AGL
Alerts are a function of flight phase (T/O or landing) based on A/C configuration
Decreasing performance--
Upper half ON--red
Aural Warning--"WINDSHEAR" 3 times
Increasing performance--lower half ON--amber
Note--*Windshear aural warnings take precedence over all other GPWS alerts, and also TCAS alerts. Other GPWS visual warnings, such as GS light, occur as normal.*

Enhanced GPWS
Modes 1 through 7—Same as above
Added Modes:
Terrain Awareness Alerting
Terrain Clearance Floor Alerting
Currently—727 EGPWS has *no* terrain display capability
Operation
Uses aircraft position and
A worldwide database of
Airports with runways over 3500' in length and
Worldwide terrain features
Uses GFMS for position input
Design terrain area—
3 Either side of airplane
± 1/8 Nautical mile either side of airplane

Warnings

Impending terrain conflicts predicted based on current aircraft flight path

Distance system "looks ahead" varies with airspeed and other flight flight parameters

Caution Alerts—Aircraft 40-60 seconds from potential terrain conflict .... *"CAUTION TERRAIN"*

Warning Alerts—20-30 seconds from potential terrain conflict........... *"TERRAIN, TERRAIN, PULL UP"* (*"PULL UP"* is repeated continuously)

Terrain Clearance Floor Alert—Vicinity of airport runway, increasing vertically with distance from runway*"TOO LOW   TERRAIN"* (*Repeated twice, sounds again for each 20% reduction in radio altitude*)

EGPWS Switch

Pressed—**OVRD** (*White*) Alerting functions disabled

Pressed—OFF (*Blank*) **OVRD** light off, switch blank, system operates normally

FAIL—(*Amber*) Fault detected in system

Traffic Alert & Collision Avoidance System (TCAS)

Altitude Band Selection

ABOVE--

Used in Climb

Traffic displayed -2700 to +8700 from present altitude

NORM--

Used in cruise

Traffic ±2700' from present altitude

BELOW--

Used in Descent

Traffic +2700 to -8700' from present altitude

TCAS Operating Policy

Both CA & FO must be qualified to operate

Visual contact made with traffic--See & Avoid

TCAS RA Issued--

MUST comply with warning, even if traffic in sight

Assures mutual collision avoidance as coordinated through TCAS occurs

Only vertical avoidance maneuvers permitted in response to RA

If in a turn , continue the turn

If "*INCREASE CLIMB*" RA occurs, level wings and climb at faster rate

Pilot Flying makes smooth, controlled responses

Assumed reaction time:

5 Seconds from RA command issuance

2.5 Seconds from reversal command for existing RA

No Enforcement action to be taken against flight crews complying with TCAS who deviate from ATC clearance

Traffic Symbols and Warnings

Other Traffic--Open Diamond .... White ◇

More than 6 miles away *OR*

Over 1200' altitude differential

Proximate Traffic--Filled Diamond .... White ◆

Within 6 Miles AND

Within ± 1200' altitude

Traffic Advisory (TA) Intruder--

Filled Circle..Amber ●

A target whose altitude is projected to be within ±900' at point of closest passage

TA Alert occurs at 40 seconds from point of projected closest passage

Aural Warning.......................... ***TRAFFIC TRAFFIC***"

Resolution Advisory--Filled Square .... Red ■

A target whose altitude is projected to be within ±600' at point of closest passage

Threat is 25 seconds from closest point of approach

Vertical avoidance maneuver displayed on IVSI

Aural Warnings--See Flight Manual p 13-Flt/I-21

Cross References

2-Lim-9	Limitations
3-App-Ldg-G/A-4	"Minimum maneuvering speeds…on Landing Data Cards…based on 1.3 times stall speed (1.3 VS) and provide adequate stall margin to at least 30 of bank plus overshoot."
3-App-Ldg-G/A-8	FE Monitoring/Callout Items
5-Air-1	E & E Cooling
5-Air-13	E & E Cooling Light

14 -- Fuel System

Fuel Use
Without Aux Tank Fuel
#2 is Fed until all three tanks equal
Feed Tank to Engine for all three tanks--remainder of flight
Aux Tank Fuel
#2 tank must have over 10,000 pounds more fuel than either #1 or #3
Use 8000 pounds from #2
Use Aux Tank fuel
Feed fuel out of #2 until #1, #2 & #3 all have the same amount
Feed Tank to Engine for all three tanks--remainder of flight

Refueling Sequence
1, 2, & 3 Are filled simultaneously
Top Off #2 last
-200B, Aux tank last
Alternate method
Uses Space left in #2 tank
Fuel pumped through #2 tank crossfeed valve and back up to aux tank
See schematic, p. 11-Fire-4

Fuel Panel Schematic Notation
Gage represents the tank
Switch represents Boost pump location
Boost Pump Low Pressure Lights represent pressure sensors
Arrows are check valves

Boost Pumps
Bypass Allows suction feed in event of BP Failure
Suction feed is supposed to be capable of providing fuel to the Engines at all altitudes up to approximately 25,000 Feet
Can't use fuel heat in conjunction with gravity feed
Pump Output--on The Order of:
20 psi--Normal
30 psi--Over pressure Boost Pump
-200Bs, #2 Tank
On the order of half again as much
(Above numbers are for awareness, and are not intended to be construed as exact)

Fuel Shutoff Valve Control
Fire handles Close fuel shut off valves
N716-N731 (Ex-Braniff)--Start Levers also close fuel shutoff valves

Fuel Temperature
Sensor Located in #1 Tank
If at or below 0ºC--Fuel pre-heat required before descent

Fuel Shutoff/Fuel Boost Pump Power Sources
Fuel Shutoffs on Hot Battery Bus
Crossfeeds--Battery Bus
RALF--Right Aft, Left Forward #2 tank boost pumps are on Essential Bus
Other Boost Pumps--

☐#1 Tank Aft	#1 AC NE
☐#1 Tank Fwd	#3 AC NE
☐#2 Tank L Aft	#2 AC NE
☐#2 Tank R Fwd	#1 AC NE
☐#3 Tank Fwd	#2 AC NE
☐#3 Tank Aft	#3 AC NE

Aux Tank
Fwd--#2 AC NE
Aft--#3 AC NE

Fuel Dumping
Dump Nozzles in both wing tips
Dump/fueling/defueling use same manifold
Fuel dumping stops at 3500 pounds per tank (except in Sim)
Dump rates
Dependent on # of pumps running
300 ppm per Boost Pump
-200B
Over pressure pumps @ 360 ppm
Slightly faster dump rate results
Dump Rate-About 2500 ppm
-200B-About 2650 ppm
Dump Shutoff Valve
Opens with:
More Than 3500 pounds in tank
At least one boost pump operating in respective tank
Prevents dumping too low
Left after dump=10,500 pounds of fuel; *all usable*
No safety interlocks to prevent ground dumping

APU Fuel Feeding
Gravity feed from #2 tank below 750 pounds static (early -200As) may result in APU flameout
Limit in #2 tank with A/C moving:
-200A up to 870AA--4800 pounds
-200A after 871AA, & all -200Bs--Pressurized fuel to APU, so no set minimum

Imbalance between #1 & #3 Tanks
Can't *move* aircraft with over 1000 pounds imbalance

FUEL

Fuel Quantity Gage Test

Runs FE gages Down, wing refueling panel gages up

Therefore, if refueling is occurring, don't test gages

To determine if fuel truck connected on PDCS-equipped aircraft:

Press Fuel Key

If error message results, fueler is connected

If total fuel quantity is displayed, fuel truck is not connected

Gages are electronic, with digital readout

Caution--*Holding test button over 30 seconds induces a "hard" error in the gage; See also page 3-13.*

Test results in cycling through the following indications:

Hold button until gage goes blank (3-4 seconds)

Gage tests all characters

Gage reads maximum tank reading

Gage reads actual fuel

All Crossfeeds Open, All Boost Pumps On (Except Aux Pumps, -200B), When:

♦ Dumping

♦ Less than 1000 pounds in a main tank

♦ Takeoff with over 178,000 Lbs gross weight

♦ Takeoff with Fuel in Aux tank

♦ Electrical Fire or Smoke of Unknown Source

Takeoff Configuration

Normal

Tank to Engine

All Boost Pumps on except Aux Tank, (-200B)

All crossfeeds *closed* except #2

Exception:

If Aux tank has any fuel *OR*

Over 178,000 Lbs Gross Weight

Then

Not Tank to Engine

All Boost Pumps on except Aux Tank, (-200B)

All crossfeeds *open*

Cross References

2-Lim-7	Limitations
3-Pre-Flight-19	Fuel Slip Allowances (±300 Gal)
3-After Ldg/Parkng-4, 8-APU-9	Min #2 Fuel Tank Quantity for APU operation
3- After Ldg/Parkng-2	Post Flight Fuel Procedures
4-Emerg-11	Loss of All Engines--Restart
14-Fuel-13	Pre-T/O Fuel Panel Configuration
14-Fuel-13	Fuel Use After T/O
14-Fuel-13	Fuel Heat Operation
20-Eng-5	Engine Fuel Lines
20-Eng-36	No Throttle Response
20-Eng-38	High Oil Temperature (Cycle Fuel Heat)
MEL 28-X	Fuel System Covered in MEL Section 28

15 -- Hydraulics

General--
Closed system--Fluid Circulates only when used
Pressure Relief
3500psi
Normal System Maximum Pressure--3175psi
Clear synthetic fluid
Turns yellow when mixed with air
Cannot be cleaned from clothes easily
Mixing with oil-based products can deteriorate seals
Dangerous in eyes
A-Pumps--Turned by Engine Gearbox
B-Pumps--Electrically Driven
A System Reservoir pressurized by Bleed Air (approx. 45-50psi)

Fluid Quantities
A System--
4.4 Gallons, full
3.5 Gallons to leave gate
3.0 Gallons to take off
2.0 Gallons--warning light on
B System--
1.8 Gallons, full
Full to take off
1.5 Gallons--warning light on
Standby System--
.52 Gallons, full
.28 Gallons to take off
No Quantity warning light

A-System

Items--NEAT GOLLFF
Nose Wheel Steering, Brakes if installed
Elevator
Aileron
Tail Skid
Ground Spoilers
Outboard Spoilers
Landing Gear
Lower Rudders
Flaps--Leading (LEDs)
Flaps--Trailing

Hydraulic Fluid Shutoff Switches
Electrically Shut off fluid to/from pump
Remaining trapped fluid lubricates pumps as they continue to turn
Pumps are turned by engines, so keep turning and generating some pressure even at windmill speeds
Shutoff Valves are Battery Operated
Closed when
Fire handle for respective engine is pulled
Guarded switch is turned off
Solenoid-type switches
Require electrical power to hold closed
Fail-safe position is open

Low Pressure Lights
On at 1200psi output
"Out to Lunch Lights"--12:00 Noon
11 total
1--Pneumatic Brakes (Capt's panel)
6--Set of 6 System Low Pressure (F/O's panel)
2--A System Circuit lights (F/E Panel)
Deactivated by pulling Fire Handle
2--B System Circuit lights (F/E Panel)

Cooling of A System Fluid
Fluid is heated as it is pressurized and used to power actuators
Cooling is accomplished in fuel tank #3, in return line
Overheat sensor is in the same return line

Low Pressure Pump Switches
Decreases Pump output to near zero
Used for leak testing & troubleshooting

B-System

Items Powered--Break In Up Stairs Every
Afternoon
Brakes
Inboard Spoilers
Upper Rudder
Stairs
Elevators
Aileron

Pump Switches--Off
Stops pump from turning
No lubrication needed, so no fluid shutoffs

Reservoir Low-Level Light--
1.5 Gal

Overheat Sensor
In case drain--not return line

Cooling of Fluid
Accomplished In Fuel Tank #1, in return line

Standby Hydraulic System

Items Powered--Low Lead
Lower Rudder driven as a backup to A System
  Power for electrically-driven pump from Essential Bus
Leading Edge Device Backup
  Power but not fluid to LEDs
  Locks out A system fluid
  Cannot retract w/o A system
  Fluid supplied from "Baby B"

Green Standby System Light
"In to Lunch" Light
On when pressure over 1200psi

Cross References

2-Lim-8   Limitations
4-Emerg-49   Manual Reversion Landing
12-Flt/C--13   Alternate Flap Operations (See also 10-Elec-8)
17-Misc-2   Warning Horns
PERF-30   Performance with Tail skid Down

Alternate Flaps

Alternate Flap Lowering
AC Motor
  provides backup to Hydraulics
  Mechanically connected to trailing edge jackscrews
Guarded Master Switch
  Arms Inboard & Outboard switches
  Cuts off System A pressure to all flap actuating devices
  Turns on Standby pump
Inboard & Outboard switches
  Open shutoff to hydraulic-driven pump for LED
  LEDs come down randomly and slowly
  TE Flaps have electrical backup (takes up to 6 times as long)
  Limit--15 degrees TE Flaps
Spoilers not held down--Will float up & create drag when standby system used to lower flaps

16 -- Landing Gear

Gear

Landing Gear
Warning Lights
Green down lock lights
Down lock switches closed
Amber/Red gear warning lights
Gear unlocked
Gear & Gear handle positions disagree
Gear Warning Horns
Steady Horn--*Always gear-related* if it's a steady tone
♦ Throttle pulled back (≈ *3 to 4 inches from aft stop*)
♦ Gear not down & locked
♦ -200B—Flaps over 27° and gear not down
♦ -200A—Flaps over 22.5° and gear not down*
Flap-related horns can't be silenced without gear down
*Note: *Inhibited when EPR above 1.48 to allow for flaps 25° takeoffs.*
Low Power Horn always active regardless of flaps:
Throttle pulled back
Cutout handle not pulled
Note: *Gear won't come up w/o squat switch in "flight" mode*

Gear Doors & Handle
Must be in agreement when pressurizing hydraulics
If not, doors will move when A System hydraulics are pressurized
Safety inspection on Origination Pre-Flight (agreement check) required for this reason
Wheel Well Gear Door Handle in DOWN position prevents door closing on ground

Tail Skid
General
Hydraulically actuated
Down when gear handle lowered
No connection to flaps
Warning light – Disagreement light;
On when skid and gear handle arenot in the same position (up or down)
Should be checked during gear cycle
Performance penalty if stuck down
Climb--25% (Distance, Time, Fuel Flow)
Cruise--10% (Fuel Flow)

Brake Systems

Brakes
Powered by
System B
System A through the interconnect on ground
Pneumatic Brake System
Use when B System pressure is not available for braking.
Cautions:
Keep feet off brake pedals
Shuttle valve could block pneumatic pressure
No differential braking
No Anti-skid

Lockout Deboost Valves
Reduce pressure from system operating to that required at the brakes
Prevent fluid leak in one brake from depleting all system fluid by isolating that brake
Anti-skid
Modulates pressure on wheels decelerating too rapidly
Releases pressure in event of a wheel lockup
Protection above 15 knots GS
Can fully or partially reduce pressure to modulate braking action

Locked Wheel Protection
Only on -100s
No longer a factor

Nosewheel Brakes
Ground speed over 15 Knots
Brakes pedals over half depressed
Nosewheel not turned more than 5°

Miscellaneous

Safety Relay Functions
Items Which Have Ground-Activated Functions (**AVA FLAGS**)
*A*nti-Icing
*V*oice Recorder
*A*PU
*F*LASS--T/O Warning
*L*anding Gear
A/C
*G*PWS
*S*tall Warning
More detail--See Page 16-2

Cross References
3-Pre-Flight-14	Gear Pin Removal (maintenance required)
3-Starting-3	Brake Pressure 2000 psi before pressurizing; Pneumatic brake pressure minimum 1200 psi
Maneuvers-3	Brake Cooling Chart (T/O Abort)
3-App-Ldg-G/A-6	Braking & Anti-Skid Ops
4-Emerg-32	Abnormal Gear Configuration
4-Emerg-30	Manual Gear Extension

17 -- Miscellaneous Systems

Instrument Power Sources & Cockpit Lighting

The following is a summary of how various degrees of power loss effect instrumentation, lighting, and communications. Some instrument power sources will vary on some aircraft. Refer to the operating manual for specific differences. Unless stated otherwise, assume that items available are cumulative; items in each category are added to those in the previous list

---

Condition: **No Generator**
                   Battery OFF

Capability: **Maintain attitude and basic aircraft control for up to a maximum of 5 minutes in IFR conditions. (Time guarantee is based on Standby Attitude Indicator)**

Instruments & Radio

Pitot/Static Instruments (No Heat)
Standby Horizon (5 Minutes)
Standby Compass

Lighting

Walkout Lights--Emergency Exit Lights, Passenger Aisle Path Lights and over wing Slide Illumination Lights (If Armed). 1st of the "4 Ws"

Condition: **No Generator**
                   Battery ON

Capability: **Maintain attitude and basic aircraft control. Duration depends on Battery charge.**

Instruments & Radio

Standby Horizon Powered by Battery

Lighting

Walnut Lights--White Background Lights on Capt & F/O Panels
White Dome Lights
Whiskey Compass Light

Condition: **No Generator**
                               Standby Selected on Essential Power Switch

Capability: **Navigate**
                               Communicate
                               Penetrate
                               Shoot an ILS
                               Still limited to the Battery life duration

Instruments & Radio

Capt HDI
        Attitude
        Runway Symbol
        Glide Slope
Capt CDI
        CDI Needle
        Card (With selection of Capt CDI on Alt)
Capt RMDI Card
F/O CDI (Card Only)
#1 VOR#
#1 VHF Nav
#1 VHF Comm

Lighting

A/C 876 and some subsequent aircraft have lighting in:
        Center (#3) Altimeter
        Center (#3) Attitude Indicator
        Center (#3) Mach/Airspeed Indicator

Condition: **Essential AC & DC From Any Main Generator**
                               Battery Switch ON

Capability: **Essentially full instrument capability, and _time_ is now available to sort out the other problem**

Instruments & Radio

Capt CDI--Command Bars from Flight
        Director
Capt Pitot Heat
#1 DME
#1 Airspeed Warning
#1 Transponder
#1 CADC--Ex Braniff aircraft
#2 CADC--AA -200s
Radar (No Stabilization)
ADF (In Both RMDIs)
        #1 Needle International
        #2 Needle Domestic
Autopilot (Only on N716-731, Ex-Braniff)

Lighting

KEEP: Walkout Lights--Emergency Exit Lights, Passenger Aisle Path
              Lights and over wing Slide Illumination Lights (If Armed).
              Standby Compass Light (Whiskey)

LOSE: Center (#3) Altimeter
              Center (#3) Attitude Indicator
              Center (#3) Mach/Airspeed Indicator
              Walnut Lights--White Background Lights on Capt & F/O
                Panels

GAIN: (3 **P**s)
              Forward Panel Background Lights (**P**eanut)
              F/E Panel Background Lights (**P**eanut)
              Cabin Lights (**P**eanut)

Note--Major limitations in this condition are no B Pumps and Manual Trim Only.

                Boeing 727 Study Guide ©1992, Updates 1993-2019

18 -- Navigation Systems

Compass Systems

DG (Directional Gyro; Heading Information)
RMDI=Radio Magnetic Deviation Indicator
CDI=Course Deviation Indicator
#1 DG--Capt RMDI, Slaved==>F/O CDI
#2 DG--F/O RMDI, Slaved==>Capt CDI

Global Navigation/Flight Management System (GFMS)

General
Uses 24-satellite "constellation" to reference position
Accuracy—± 50 feet
Uses four satellites to cross-measure position
With only three satellites, can operate at reduced accuracy using aircraft altimeter input

Receiver Autonomous Integrity Monitoring (RAIM)
Fault detection programming
Monitors accuracy and annunciates errors
Measures the four navigation axes continuously
    Four satellites *or*
    Three satellites and aircraft altimeter

Fault Detection and Exclusion (FDE)
System adds additional satellite measurement up to 6 total
Detected inaccuracies result in isolation of bad satellite signal and exclusion from navigation computations

Required Navigation Performance (RNP)/Actual Navigation Performance (ANP)
RNP—Accuracy measure in NM required for various flight regimes
Five default values programmed:

Approach:	0.5 NM
Terminal area:	1.0 NM
Takeoff:	1.0 NM
Enroute, Domestic:	2.0 NM
Enroute, Oceanic/Remote:	12.0 NM

Values above may be overwritten manually and will be annunciated as errors from entered value

Operation—See Section 18 (Navigation) pp. 56-98

19 -- Oxygen System

Aircraft Fixed Systems

Cockpit System
- 1 Cylinder
- Minimum pressure at @70°F, Domestic or Extended Over Water
  - 1200psi Cockpit
  - 1500psi Cabin

Cabin System
- 2 Cylinders
- Minimum pressure--1500psi,

Green Disk (external) ruptures for over pressures in either system (Oxygen Discharged Overboard)

Pressure gages--battery bus

Portable Systems

Portable Cylinders & Regulators
- All in cabin
- MEL SYS 35 covers required numbers

2 Regulator Connectors
- Green--normal
- Red is therapeutic--twice the flow

Pilots Use of Oxygen Required

Cabin over 10,000' pressure altitude, all crew on duty with masks on

A/C over 25,000, if one pilot leaves station, other must wear mask

A/C over 41,000, one pilot must wear mask at all times

Automatic Activation of Cabin Oxygen System

Two Redundant & Independent Switches
- Electro-pneumatic
- Manual-pneumatic

Both are set to activate at Cabin altitudes of 14,000'

3 Ways to drop Passenger Masks

Electrically

Manually

Pneumatically (Also called *Accidentally*--By letting cabin get above 14,000'!)

Passenger Oxygen System

Power Source Battery Bus For:
- Passenger oxygen system Gages, Switches, and Light

Passenger Oxygen Light
- ON--Manifold is pressurized
- Also should light "No Smoking" (NS) and "Fasten Seat Belts" (FSB) lights

Actions at Activation of Cabin Oxygen System
- Lights on in cockpit
- "No Smoking" & "Fasten Seat Belt" lights on in cabin
- Pressure opens PSUs

Reset handle

Closes both electro- and manual-pneumatic valves

Must be safetied Up *and* Down--should be in the center position to prevent being locked in the manual reset position

PBE (Protective Breathing Equipment)

One in Cockpit

Three in Cabin

On origination pre-flight, check:
- PBEs Present & Condition
- Integrity of seals on PBE containers

Cockpit PBE seal (if PBE present) also checked on through-flight inspections (Requirement listed on checklist card, not in flight manual)

Cross References

2-Lim-10	Limitations for Bottle Pressures
2-Lim-15	Restowing Authorization for F/E (In Flight ONLY)
3-Starting-1	Preflight of Oxygen System
4-Emerg-5	Use of Passenger Oxygen System below 14,000' not recommended with smoke in the passenger cabin; it won't help passengers filter out smoke, since O2 is mixed with ambient air; may increase fire hazard by increasing O2 saturation in cabin
4-Emerg-19	Explosive Depressurization
4-Emerg-41	Location of Equipment

20 -- Power Plant

General-- JT-8D Engines

Aircraft/Engine Mix

A/C	Primary Engine	Alternate Engine
-200A	-9A	-15A
-200B	-15A	-9A

Accessories Driven by N2 Compressor Accessory Drive (SHAFFTO)

Starter
Hydraulic Pump (#1 & #2 Engines)
AC generator
Fuel Control
Fuel boost pump (Engine-Driven Fuel Pump)
Tach generator
Oil Pump

Bleeds

N_1 Bleed--Pre-Cooler

N_2 Bleeds--

From 6th, 8th, & 13th stages
Fuel Heater uses 13th stage air
Other details--see Section 5 Notes

Engine Bleed Air Uses--(WEE FACT)

Wing Anti-Ice
Engine Anti-Ice
Engine Start
Fuel Heat
Air Conditioning & Pressurization
CSD Oil Cooler
Thrust Reversers

Gages & indicators

Warning Lights

#2 Engine Inlet Duct Access Door Light
Indicates #2 engine inlet access door latch in the aft stair area is unlatched
Fire *Detection*--Essential
Except Fire detection on **APU** is Battery
Fire *Protection*--Battery
Except Fire discharge **Bottle Lights** are non-essential DC
Low Oil pressure Light
Below 35 psi
Sensed by separate sensor from gage
Oil Filter bypass light
Comes on at lower value than what is required to bypass filter
Provides warning of impending clog
Fuel Icing Light—Acts as Fuel Filter Bypass Lights
Come on when differential pressure across filter is sensed
Impending filter bypass is warned
One Light for each engine
If on, and temp ≤0°C, apply heat
Lights on oil and fuel system are *__not__* a positive indication that bypass is occurring--only that it should be
Watch oil pressure & fuel flow to ensure continuing flow

Triggered by pressure differential on either side of filter
If lights come on and temperature is >0°C, cause is likely mechanical blockage, not ice

Gages

N_1, N_2, EGT--self powered
All FE gages are electrical
Fuel Flow Gages--
AC/DC--Freeze after power loss
Oil Quantity (FE Panel)
AC--Freeze after power loss
Oil Pressure
40-55 psi
Yellow arc 35-40
Separate sensor keys light on Pilots' panel
Fuel Temperature Gage
Sensed in #1 (L wing) tank--worst (coldest) case
A system hydraulics cooled in #3 Tank
B System hydraulics cooled in #1 Tank (Lower load, so cooler)
#2 Tank in fuselage--warmer than wings
EPR Gage measures {Station 7}/{Station 2}
Fuel Flow--AC+DC inputs

Throttle-Activated Takeoff Warning Horn (FLASS)

Warning horn sounds if any of the following are not in the specified position:
Flaps--Takeoff setting of 5°, 15° or 25°
Leading Edge Devices--Down, Green light ON
APU--Off
Speed Brake Handle Forward
Stab Trim in the Green Range

Fuel Control

Cable Linkage from throttle to fuel control

Start Levers

Cutoff
Ignition off
Idle
Ignition on
High intensity ignition provided by battery
Starting Ignition *terminates* when start switch released

Ignition

Starting--High intensity ignition provided by battery
Continuous--Low intensity ignition is on 115V non-essential AC
Switches
Flight
High intensity ignition
Off, with Continuous ignition ON
Low intensity ignition
Ground
High intensity ignition--occurs with start lever in idle
Start Valve opens

Continuous Ignition Req'd for: (**TWEETCAT VH**)

Takeoff
Wing Anti-Icing On
Engine Anti-Icing On
Emergency Descent
Turbulence
Compressor Stall
Approach & Landing
Training & Test Flights
Volcanic Ash
Heavy Precipitation (Flight Position)

Overhead Panel Switches

Start Switches
Start Valve Open Lights
Continuous Ignition Switch
Engine Door Access Light

Oil Quantity

Minimum Quantity if next day's flight segment unknown, 2.5 Gallons (3/46b)
Standard oil type changes periodically; See Operations Manual for current guidance
Same type oil is used for:

- Engines   - CSD
- APU   - Starter

Oil Quantity Test Button--Decreases Gage Reading, then back up to sensed quantity
See Section 2 for pressure and temperature limitations

Fuel Anti-icing

Uses 13th stage air to pre-heat fuel enroute to fuel filter
Turn on for 1 minute with 1500 PPH fuel flow
Rise in oil temp is the only positive indication of application of fuel heat
Maintain 1500 PPH fuel flow for 2 minutes cooling after fuel heat is turned off
Used for situations of fuel temp $\leq 0°$ *and*:
Icing light on
Do affected engine first
Then do others
Prior to approach and landing by 5-6 minutes
Do #2 first (first engine to be pulled back)
Then either #1 or #3
Before Takeoff
All three at once, for one minute
No cool down period required
Two factors minimize danger of vaporized fuel during ground operations:
Lower RPM, and lower bleed air temperatures
Higher ambient air pressure

Reversers

Light indicates not locked in forward thrust position
Ground static test--80% N_1 is maximum
Power Back--80% N_2
Landing
84% N_1 on the runway for after landing deceleration
F/E call out is required at 80% N_1
At low Forward speeds, danger of FOD ingestion is increased

N_1/N_2 Limits

On the flight side of the hold short line, most RPM limits N_1
On the ramp side of the hold short line, most RPM limits apply to N_2

Start

Light off within 20 Seconds
Duct pressure below 25 psi—F/E notifies pilots
Fuel flow up to max 1500 Lbs, then back down
Oil pressure on Cold soaked engine (Below 25°C)
Can stagnate
Oil Filter Bypass light can be on for up to 90 seconds
Oil low pressure may be below normal (40 psi) for up to 5 minutes on start (Conditionals, page 3A-Cond-17)

Fire Warning

Fire Sensor Loops

Loops on all three engines
Loop in wheel wells (1 light only)
Test function only tests circuit continuity

Power Sources

Hot battery bus
Fuel shutoff switches ....Fluid
Battery bus...⇩
Engine hydraulic fluid shutoffs ....⇩
Generator Field Relay...Elect
Freon bottle arming/Activation........................... ⇩
Essential AC
Fire detection system ....⇩
Non-Essential AC
Bleed Valves.. Air
Engine anti-ice ....⇩

Freon bottles

Button
Discharges bottle
Aims freon to engine with handle out
3 Disks on Right side under pod
2 Red disks -- thermal over pressure discharge
Yellow disk -- intentional discharge of either bottle
Pressures--see limits

Pulling Fire Handle:

Isolates engine from fuel, hydraulics, and unnecessary bleed air drain
2 Fluids
Fuel cut at engine fuel shutoff valve
Hydraulics
A System pump fluid shut off (except #3 engine)
Disarms A System low pressure light
2 Electrics
Generator field Opens after 5-10 sec.
Arms Freon Bottle Discharge Switch
2 Airs
Bleed air shutoff
Same effect as turning off engine bleed switch(es) for respective engine
Closes Duct Isolation Valve (If installed)

Anti-ice external to engine

Wing Anti-Ice Valves Closed (1 & 3 only)

Closes #2 engine Inlet Duct Anti-Icing

Does not effect Engine anti-ice for respective engine, or for #2 Inlet

Cowl (pod engines) & compressor face (all 3 engines) stays on

Engine Failure to Respond to Throttle

Two conditions can result in an:

Engine changing to or remaining at high thrust *with*

No movement of throttle *and*

No engine response to throttle movement

The two conditions are:

◆Complete loss of N_2 signal to the fuel control unit (FCU)

Results in FCU setting thrust between 92-95% N_2 regardless of throttle movement

◆Partial loss of N_2 signal to the FCU

Caused by wear and slippage of the N_2 shaft

FCU Interprets as lower than commanded N_2 and compensates with more fuel and can result in high EGT

Throttle position cannot control engine in either case

FCU response designed to protect high thrust in critical situations

Performance Data Computer System (PDCS)

Computes optimum engine performance for various flight modes

Should be used "at every possible opportunity"

May not be used when:

Mixed engine configurations

EPR Gage inoperative

Fuel quandity gage inoperative (limited use see PDCS abnormal operation procedure No. 2)

On takeoff, -200A aircraft

Raisbeck Hushkits

Allows –200As to comply with noise restrictions when modified

Components

Specialized inlets

Extended and redesigned tail pipe

Combined with other system & operational modifications

Slightly reduced travel on leading edge devices

Reduced wing camber causes lower drag and requires less thrust

Approach V Speeds increased approximately 4 knots

Landing flaps limited to 25 except for emergency

Landing configuration warning sensors repositioned

Cross References

2-Lim-10	Engine Limitations (Through 2-12)
3, Starting-7	Engine Start Sequence; Last engine not less than 3 minutes prior to takeoff
3-Starting-13,14	Engine Start Procedure
3-Starting-15	Interrupted Engine Starts
3-Starting-15	Start Valve Motoring Cold Soaked Engine--up to 4 minutes to open valve, then 1 minute after open
3-Taxi-Takeoff-3	Setting Takeoff Power
3-After Landing-8	Minimum Oil after shut down
3A-Cond-17,19	Fuel Heat Requirements
3A-Cond-23	Taxi Out with tight gate positions maximum N_2 68% (1 & 3) above 70°F, 65% below 70°F
3A-Cond-28	Engine Spool Up time Approx. 8 Seconds
5-Air-13	Engine Fail Warning & Auto Pack Trip
5-Air-29	Auto Pack Trip Fail To Arm Procedures
11-Fire-5	Engine Fire Handle Functions

Key Acronyms & Memory Items

All of the lists found in this section are found under their respective chapter headings as well. They are consolidated here for ease of study.

A/C & Pressurization

Auto Pack Trip Armed when: (GAFT)

Ground-- A/C on Ground
Auto Pack Trip Switch NORMAL
Flaps not up
Throttles all above 1.5 EPR
    Has separate sensor--doesn't rely on panel gage

Aft Zone Temperature Switch Four things needed for proper operation--(VASS)

Valve--Air Mix Valve About 1/4 travel left toward the cold side (for colder in cooler in rear only)
Automatic on temperature rheostat
Sidewall--Cabin Air distribution lever in sidewall
Stabilized Flight (cruise)

Functions of the Outflow Valve -- NARC

Negative Pressure relief
    Set at -1.0 psi
Altitude limiter--Set at:
    13,000' ± 1500', Pneumatic System
    15,000', Electronic System
Relief of excess Pressure (Pressure relief)
    Occurs in 8.9-9.6 psi range
    Dumps to 8.0 psi rapidly
Cabin pressure regulation

Key Altitudes & Pressures

10,000'
    --Warning Horn
        Note--*Also raises the PA Volume*
        Resets Descending the cabin below 9,500'
    --Max dispatch altitude with known press problem
13,000' ± 1500'--Altitude limiter--Pneumatic pressurization
14,000'
    --Emergency O_2 System Activation (pneumatic and electrical switches)
    --Electronic Pressurization switches from AUTOMATIC to STANDBY Mode
    --Max altitude after depressurization
15,000'--Outflow Valves Close--Electronic pressurization
8.6psi--Auto Controller Limit (±0.15)
9.6psi--Outflow Valves open in a range from 8.9-9.6psi; dump down to 8.0psi

Autopilot

Interlocks--Aileron Channel

Prevent Engaging *(EMPTY)*:
    Mode Selector **not** *in* **M**anual
    Turn Knob **not** *centered*
Disengage **OR** prevent Engagement
    Put **D**own **Y**our *(EMPTY)* **V**anilla **C**one
    Power--*lost on No. 2 Radio Bus (ESS, Brannif)*
    Disconnect Button on Control Wheel
    Yaw Dampers--*Both Off*
    VG--*Loss of power or fast erect cycle entered*
    CDI Select Switch moved out of Normal

Interlocks--Elevator Channel

Prevent Engaging *or* Disengage *(ACTS)*:
    Aileron Channel off
    Cruise Trim
        Cutout switch to Cutout
        Cruise Stab Trim switch moved
    Trim Switch Moved ("Pickle" switch on control yoke)
    Servo Switched (A to B, etc.)

Electrical System

Hot Battery Bus--(BAR BBGH)

Battery Bus
APU Cranking & Starting
Relay Control
Battery voltmeter & Switch
Bus Protection Panel
Generator control Panel
Hot Battery Transfer bus

White CBs (VET)

Voice Recorder CB
Essential Bus Tie CB
Trim
    Main Electric (Fast) Trim CB
    Cruise & Autopilot (Slow) Trim CB

Sync Bus Tie Automatic Shutdown--Exciter Problems (SEEP)

Stability Protection
Excitation--*Over or under excitation*
Exciter Ceiling Protection (not cleared by other protective circuits)
Phase imbalance will trip all three

Generator GB Automatic Shutdown--Speed Problems (FACED)

Field Relay of Generator Tripped
APU GB Closed
CSD overspeed or under speed
External Power on
Disconnect of CSD

Generator Field GB Automatic Shutdown--Voltage Problems (VDF SEE)

Voltage--Over or under voltage situation
Differential Fault--Feeder shorted or grounded
Fire Handle Pulled
SEE--From BTB, **SEE** items will trip Field if problem persists after BTB opened

APU Generator Breaker Automatic Trips (APU'S Field)

Alternate source powering Sync Bus
Phase **U**nbalance with APU powering Sync Bus
Speed Faults
APU **Field** Relay Tripped

Ground Interconnect Electrical Switch

Items with a Ground-Only Function (**SYRIA**)
Stall Warning
  Deactivated
Yaw Dampers
  Deactivated
Recorder (flight)
  Deactivated
Interconnect (Ground)
  Activated--Hydraulic Ground Interconnect
  Needed on ground for hydraulic tests
Autopilot--Deactivated
Also--Fueling Valves Operational

Lighting ON--Battery Power Only (Four W's)

Walnut Lights--Emergency Background Panel lights
Whiskey Compass
Walkout--Emergency Exit Lights
White Dome Lights

Fire Protection

Pulling APU Fire Handle (BIG FAT)

Last 3 immediate; first 3 several sec.
BLEED VALVE--APU Bleed Valve closes as airflow ceases
ISOLATE--Isolates APU in shroud
  Spring-loaded Cooling air discharge valve closes as airflow decreases
GEN--APU Generator falls off line as APU spins down

FUEL--Closes APU fuel shutoff valve
ARM--Arms APU Freon bottle
TRIP--Trips Generator Field

Pulling Fire Handle:

Isolates engine from fuel, hydraulics, and unnecessary bleed air drain
2 Fluids
  Fuel cut at engine fuel shutoff valve
  Hydraulics
    A System pump fluid shut off (except #3 engine)
    Disarms low pressure light on hydraulic system panel
2 Electrics
  Generator field Opens after 5-10 sec.
  Arms Freon Bottle Discharge Switch
2 Airs
  Bleed air shutoff
    Same effect as turning off engine bleed switch(es) on A/C & pressurization panel for respective engine
    Closes Duct Isolation Valve (If Installed)
  Anti-ice external to engine
    Wing Anti-Ice Valves Closed (1 & 3 only)
    Closes #2 engine Inlet Duct Anti-Icing
    Does not effect Engine anti-ice for respective engine, or for #2 Cowl
    Cowl (pod engines) & compressor face (all 3 engines) stays on

Flight Controls

Inboard Flap Switches (PARIS)

Pack Fans
Auto Pack Trip
Rudder Load Limiter
In flight Warning of Speed Brake+Flaps
Stall Warning

Outboard Flap Switches (GOLF)

GPWS
Outboard Aileron Lockout
LED Sequence Valve
Flap+Gear Warning Horn

Hydraulics

A System Items--NEAT GOLLFF

Nose Wheel Steering, Brakes if installed
Elevator
Aileron
Tail Skid
Ground Spoilers
Outboard Spoilers
Lower Rudders
Landing Gear
Flaps--Leading (LEDs)
Flaps--Trailing

B System Items Powered--Break In Up Stairs Every Afternoon

Brakes
Inboard Spoilers
Upper Rudder
Stairs
Elevators
Aileron

Standby System Items Powered--Low Lead

Lower Rudder driven as a backup to A System
Leading Edge Device Backup

Engines

Throttle-Activated Takeoff Warning Horn (FLASS)

Warning horn sounds if any of the following are not in the specified position:

Flaps--Takeoff setting of 5º, 15º or 25º
Leading Edge Devices--Down, Green light ON
APU--Off
Speed Brake Handle Forward
Stab Trim in the Green Range

Continuous Ignition Required for: (TWEETCAT VH)

Takeoff
Wing Anti-Icing On
Engine Anti-Icing On
Emergency Descent
Turbulence
Compressor Stall
Approach & Landing
Training & Test Flights
Volcanic Ash
Heavy Precipitation (Flight Position)

Flight Data Forms

Instructions

The form is designed to record data from one flight segment. Departure and destination field data, and scheduled times can be filled out before flight. The rest is filled out when the Flight Plan/Release and Departure Plan are available. This sheet can be placed in a plastic document protector so that it is displayed below the half-sheet formatted forms. F/Os and Captains desiring to use the form will find it can be satisfactorily shrunk on a copier for use on side window clips.

Flight Plan Data: As the title implies, all data may be obtained from the flight plan. The data selected for recording is that required for programming the PDCS, ACARS, pressurization, and for determining programmed off and on times. The two exceptions are zero fuel weight and takeoff gross weight. These should be left blank until the closeout is received.

Departure Plan Data: This column is filled out from the Departure Plan, Fuel Panel, and Oil Quantity gages. Filling it out during pre-flight will remind the new (or old) F/E to add up total fuel quantity before it is called for in the Before Starting Checklist 5 minutes prior to start. It also reminds him to check oil quantity before and (hopefully) after flight to note consumption.

Departure and Destination Field Data: These columns are filled out before flight with data from Part 2 or from the communications bulletin distributed by the flight department. Frequencies then are readily available for use without having to look them up in flight.

Times: This column is used to record OUT, OFF, ON, and IN times. The boxed items are those placed on the top of the flight plan and filled in after takeoff. Ground delays are accounted for as follows:

ETO Update Entry: Required if actual ground time during taxi-out exceeds planned ground taxi time plus 15 minutes. If this time is exceeded, an ETO update is required.

Scheduled OFF Time + Estimated Time En Route (ETE) = Scheduled ON Time. If landing occurs after this time, some sort of en route delays existed and should be accounted for.

AWG Update Entry: An awaiting gate entry is required if ground taxi to gate time exceeds planned ground taxi time plus 10 minutes.

F/E Awareness Items: This section includes the IFR approach items F/Es need to be aware of and watching for. Writing these items down when reviewing the approach can help both in remembering to look at them and in documenting the items were reviewed. Writing them down also helps an F/E remember them, since he normally doesn't have direct access to the approach during its accomplishment.

The bottom of this page compiles all the checks and ACARS entries required during the course of a normal flight which are not included on the aircraft Normal Procedures Checklists. Unless things go non-standard, most flights can be completed with these entries, plus any delay codes appropriate. *Italicized items* are not required on all legs. Some of them, for instance, are required only on origination flights such as the ACARS Link Test.

Record Keeping

Many of the "Old Timers" who have been around a few years recommend keeping a record of the above items. An excellent way to do that is to keep a monthly folder for each month with an HI1 printout for the completed month, an HI3 and NS printout for each sequence, and the data sheets above. If ever called on to explain a delay code, aircraft write-up or other question, the data sheets are one way to differentiate the flight from Peoria from the one to Palm Springs. Good Luck!

GROUND CHECKS
  ACARS
    If Required:
    Link Test (1st Flight)
    Clock Chk ....<Misc>8
    Clock Blank
    Flight Info
      ATIS.............<Misc>3
      Clearance....<Misc>4
      Flight Number
      Departure Station
      Destination Station
      Fuel (AFTER Release
        Fuel Verified)
      APU Hour Meter ....
        ....<Misc>35
    ------*<Send>*------
    Load Cls Req-<Misc>5
    =======OUT Event=====
    OUT Sent
    TPS...................<Misc>12

  Flight Plan
    Times at Top
    Tracking Lines for
      Time/Fuel
    Fuel Estimate

PDCS
  <TO pg 1/2>(Rotary Sw)
  <Load pg 1/2>(Keypad)
    Forecast OAT (TPS)
    Dest Field Elev.
    Res + Alt (Flight Plan)
    ZFW(TemporaryTPS)
  <Load 2> (Page ➡)
    Cost Index (CI--Flight
      Plan)
    Flt Pln CI<60--Set 60
    Flt Pln CI>60--Set Flt
      Pln CI
  <TO pg 1/2><RCL>
    <RCL>
    Chk MAX EPRs ± .01
      from Dep Plan
  <TO pg 2/2> (Page ➡)
    Assumed Temp (TPS)
    Chk STD EPRs ± .01
    <Engage>
  <Load pg 1/2>
    Actual OAT (ATIS)
  =====Closeout Received===
    Dump Time
    Mode A/B
    ZFW in PDCS
  <TO 1>--RCL
    Max/Reduced <ENG>
    Display MAX (Page
      ➡)

DEPARTURE
  10,000 AFL--FL180
    Times @ Top of FP
  FL180--Cruise
    Departure Rep
      TO Power ....<Misc>9
        1=MAX
        2=Standard
    Delay Codes
    F/O Landing Credit
      F/O............ <Misc>26
      CNX .... <Misc>28
  VHF 1=ATIS

CRUISE
  Cooling Doors to ¼ or A/R
  Pressurization Rate Full Incr.
  Eng Mon Log-<Misc>30
  Log Book--Complete if able
  {#1 Flight} Eng AI Test .....-40
  Fuel Heat ....-40
  Pressurization Pad ....-40
  ATIS ....<Misc>3; -35
  Dest Weather.......... <Misc>7
  CHO..................<Misc>6; -30
  Notify F/As.......................-30
  Call in A/C Writeups to MX
  Review Airfield Diagram
    Taxi Distance
    Runways to be crossed

DESCENT
  Pressurization
    Bleeds
    Cooling Doors
    Cabin Pressure↘
    Temperatures
  VHF 1--Ramp
  G/A EPR's by 10,000' AGL
  "F/As Prepare for Landing:"
    "No Smoking" Tab
  VHF XMIT to #1

TAXI IN
  Voice Mode/ Monitor #1
  Call Ramp
  Approaching Gate:
  "F/As Prepare for Arrival"
    ACARS--Data Mode
    One Pack OFF
    F/O Landing.... <Misc>26

GATE
  Oil Levels
  Arrival Report:
    In Sent
    Delay Codes
    APU Hour Meter<Misc>35
    Arrival Fuel ***(Last)***

Anti-Icing Operations

Icing Conditions

Ground–8°C or 46°F OAT and:
- Visible moisture, including: Clouds, Fog with 1 mile or less visibility, rain, sleet, snow...
- Surface moisture: Standing Water, Slush, etc.

Flight–10°C or 50°F TAT and Visible moisture, as above

Exterior Preflight

Typical Ice accumulation spots
- Pack Cooling Door area
- MLG Deflectors ("mud flaps")
- Flaps & Flap hinge area rear of MLG
- Landing gear wells
- Leading Edge Devices & Top of Wings

Pay special attention to:
- Static vents
- Lavatory drains for ice accumulation
- Pressurization outflow valve area clear of ice
- Flight Controls and spoilers

Caution: *Don't power up or move flight controls if ice could obstruct movement.*

Interior Preflight

Motor engine start valves if A/C is cold soaked
- Up to 4 minutes to open
- Up to one minute to dry out after start valve light indicates valve has opened

Coordinate deicing requirements with local operations as early as possible as ice found

Station Deicing Report–ACARS <MISC> 64

Caution: If frozen contamination in intakes, must be removed by certified de-icing personnel.

Before Starting Engines

Packs -- Off for deicing
Spoilers closed for deicing (prevents re-freezing moisture from damaging spoilers & actuators)
Deicing at gate–Trim full nose down (≈ 1.5 units)

After Engine Start—

Engine Anti-Ice ON if required
Move flight control surfaces through full range to ensure clear of ice
Verify trim movement

Before Takeoff

Fuel Heat
Pre-heat Wings
Captain determines number of cycles
Flaps
- Don't move mechanical slide until flaps verified at Takeoff setting
- Retracted until reaching runway

Before Takeoff (cont.)

Move all flight controls through full range
Visually inspect surfaces
Cooling Doors CLOSED (takes 45 Sec.)

If deiced:
- Packs off until 1 minute after deicing complete
- Maximum power takeoff required

If takeoff not made within 10 minutes of starting:
- Run up engines (one at a time if necessary) to 75% N₂ "momentarily"
- Repeat every 10 minutes for taxi if in icing conditions
- Same requirements for taxi if in icing conditions

After Deicing—Mechanical Before T/O Checklist must be done in its entirety

Takeoff

Engine A/I ON by 20 Sec to power up
Engine Instruments—Ensure parameters are reasonable
Standard power takeoff not authorized:
- Deicing Fluid applied to Aircraft
- Engine Anti-Ice ON

OAT Less Than
- **-8°F**, -9A, -7°F, -15/15A
(Std Power won't Arm TO Warning)

After Takeoff

Packs OFF if fluid was sprayed in engine inlet
0'--Pack Doors Open Immediately
As Required--Cycle Gear if slush on Runway or leave down to clear
400'--Auto Pack Trip
1000' Wing Anti-Ice for
- --1 Minute--Out of icing conditions
- --Duration--In icing conditions

Pack Reinstatement if Required

Descent

Fuel Heat
Window Heat **ON**

Landing

Note--Rudder control decreases with too much reverse thrust
Over 2000' MSL--See 3-41 Pack Off Landing Procedure

	ON	OFF
Gasper	One Pack	
1000'--	Cargo Heat Outflow **CLOSE**	
	Other Pack	**OFF**
300'--	Both Packs	**ON**

After Landing

Leave flaps at 25° if Ice/Slush at Arrival Airport
Engine Anti-Ice as required
Get qualified maintenance personnel to inspect aircraft after landing

---

Flight Data Sheet

Aircraft #_____   Date: _____, 19___

Flight Plan Data

FLT # _____
FL: _____
ZFW: _____
    Closeout

C Indx _____
RSV _____
ALTN: _____
R+A: _____
RLS: _____
MACH: _____
ETE: _____
GND: _____/_____

Landing Weight

TOGW: _____
    Closeout
-FUBN: _____
=Ldg Wt: _____
V_REF: _____

Departure Plan

Fcst Temp:_____ °F
Ass'd Tmp:_____ °F
ZFW: _____
    Dep Plan

Panel

- F #1 _____
- U #2 _____
- E #3 _____
- L Total: _____

- O #1 _____/_____
- I #2 _____/_____
- L #3 _____/_____

Departure Field

Alt: _____
Conv: _____
ATIS: _____
Closeout: _____
Dispatch: _____
Maint: _____
Ramp: _____
Ground: _____
Notes: _____

Arrival Field

Alt: _____
Conv: _____
ATIS: _____
Closeout: _____
Dispatch: _____
Maint: _____
Ramp: _____
Ground: _____
Notes: _____

Schedule & Timing

SCH OUT: _____
OUT: _____
ETO-<DLA 1>: Out + 15 + Gnd:
OFF: _____
ETE
EON:
40 Prior:
ON:
AWG--<DLA 4>: ON+ 10+GND _____
IN:
SCH IN:

FE Awareness Items

Approach: _____
IAF: _____
FAF: _____
MDA: _____
Terrain: _____
Time: _____
Marker: _____
MDA: _____
Missed Approach

APU Hour Meter

Before: _____
After: _____

Final Fuel:

Notes:

Delay Codes

___ ___
___ ___
___ ___
___ ___

Titles Currently Available
By Rick Townsend

PilotStudyGuide.com

Boeing 737 Study Guide 2019 Edition
RICK TOWNSEND
Covering the 737-800 & 737-MAX Versions

Boeing 757-767 Study Guide 2019 Edition
Covering the 757-200 & 767-300 Versions

Boeing 727 Study Guide

Fokker 100 Study Guide
Includes:
Limitations
Study Notes on Each Aircraft System
Icing Conditions Checklist

Out of Print, but available by request. Contact us at www.pilotstudyguide.com

737, 757/767, 777 & MD-80 Available on Amazon & Kindle!

Boeing 777 Study Guide 2019 Edition
Covering the 777-200 & 777-300 Versions

Available only to American Airlines employees—Contact Crew Outfitters for details

MD-80 Available on Amazon & Kindle!

Sabre SYSTEM USER'S GUIDE
EGAD! WITH THESE CODES, I COULD HAVE LEFT MARS AND BEEN HOME BY NOW!
Sabre
For Sale Only to American Airlines Employees
By Rick Townsend

Douglas-Boeing MD-80 Study Guide 2019 Edition
Covering the MD-82 & MD-83 Versions

To purchase any of these titles, go to Crew Outfitters locations at the Flight Academy Gift Shop, or at DFW Airport Terminal C Operations
To purchase online, go to *www.crewoutfitters.com*

To order by phone, call **800-874-1204**

This book is available through Amazon in both Kindle and in print format!

Giving Back

Kenya Kids Can!

Feeding and teaching world changers of tomorrow

Kenya Kids Can is a non-profit working in the Rift Valley of Kenya. They feed students and construct solar powered computer centers to improve education and bring hope to needy students. Please browse the website to learn about what can be done to truly change lives.

www.kenyakidscan.org

Snowball Express

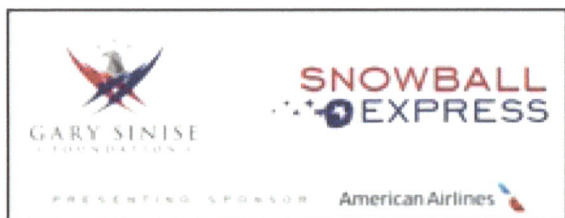

Serving the *children* of our fallen military heroes
Snowball Express Could Use Your Help!

Snowball Express honors America's fallen military service members who have made the ultimate sacrifice since 9/11 by

- Humbly serving the families they left behind
- Championing their children's future success by creating opportunities for joy, friendship, and communal healing
- Connecting these families to one another.

Since 2006, the mission of Snowball Express has been a simple, yet profoundly important one: Providing hope and new happy memories to the children of military fallen heroes who have died while on active duty since 9/11. In December each year we bring children together from all over the world for a four-day experience filled with fun activities, like sporting events, dances, amusement parks and more.

Nationally, Snowball Express provides comprehensive support programs for fallen families that are focused on transition and connections to community resources, healing and wellness, peer engagement, education and personal/professional development programs.

www.snowballexpress.org

www.ingramcontent.com/pod-product-compliance
Lightning Source LLC
Chambersburg PA
CBHW061059090426
42742CB00003B/99